JOHN FRANK STEVENS

American Trailblazer

John Frank Stevens.
*Unless otherwise noted, all photos courtesy of
John F. Stevens Papers, Georgetown University Library,
Special Collections Division, Washington, D.C.*

JOHN FRANK STEVENS

American Trailblazer

* * *

by

ODIN BAUGH

THE ARTHUR H. CLARK COMPANY

Spokane, Washington

2005

The Arthur H. Clark Company
P.O. Box 14707
Spokane, WA 99214

ISBN 0-87062-337-0
Library of Congress Catalog Card Number 2005001595

Library of Congress Cataloging-in-Publication Data

Baugh, Odin A.
  John Frank Stevens : American trailblazer / by Odin
Baugh.
    p. cm.
  Includes bibliographical references and index.
  ISBN 0-87062-337-0 (hardcover : alk. paper)
  1. Stevens, John F. (John Frank), b. 1853. 2. Railroad
engineers—United States—Biography. 3. Railroad
engineering—History. I. Title.

TF140.S755B38 2005
625.1'0092—dc22

2005001595

TABLE OF CONTENTS

Illustrations

Maps

PREFACE

I first encountered the name of John Frank Stevens when I was writing a short history about the Great Northern Railway's terminal, Appleyard, located a mile south of Wenatchee, Washington. Having worked three years for that railroad, and both my parents coming from railway families, I was interested in learning more about him. My research produced some gratifying knowledge. Stevens had also worked for the Great Northern and discovered Marias Pass across the Rocky Mountains in Montana for them. I often traveled over that route, not realizing who had found it. The same is true about Stevens Pass over the Cascade Mountains in Washington State, situated about twenty-five miles from where I lived. As I learned about his accomplishments on national and international railroads (in Canada and Russia), and the Panama Canal, I desired to write his biography.

I was discussing Stevens' activities one time with a college history professor and mentioned the discovery of Marias Pass. He replied, "Let's see, that was found by Washington State's first governor, Isaac Stevens, wasn't it?" "No," I replied, "by civil engineer John F. Stevens." It's surprising that a number of people when asked, "Who designed and planned the Panama Canal?" respond, "General George W. Goethals." Again, it was John F. Stevens.

Stevens was never honored by President Theodore Roosevelt for his work on the canal, nor did President Woodrow Wilson show any appreciation for the engineer's eight years of tremendous labor as head of the Russian railroads during World War I. It seems that America's greatest civil engineer also has been overlooked by the present generation. Yet no other civil engineer has achieved the success, been so highly honored nationally and internationally, and accomplished so much for humanity as he.

As you read this book, I hope you will appreciate his greatness and his remarkable activities.

ODIN BAUGH

ACKNOWLEDGEMENTS

I express appreciation to the following who helped make this book possible: Mrs. William Goettner, née Patricia Smith, who corrected spelling and grammatical errors and gave suggestions for changes to make the book more readable; Mrs. Lee Hyde, née Karen Marcus, who labored hours over the computer printing of the book; my wife, Nearine, who gave encouragement and endured time alone while I concentrated on writing; and the Georgetown University Library, Washington, D.C., for letting me research the John F. Stevens papers and allowing me to print some of the photographs.

I

Formative Years

Builder of railroads throughout the United States, Canada, and Russia, designer and constructor of the Panama Canal, discoverer of mountain passes, and advisor to railway officials—all of these accomplishments give evidence that John Frank Stevens, for his time, was one of the greatest civil engineers in the world. There have been other capable ones, such as the French vicomte Ferdinand de Lesseps, who built the Egyptian Suez Canal and was the initial chief engineer of the Panama Canal, but they did not accomplish the achievements of John Stevens.

The events of Stevens' life are written in massive bridges that span the abysmal depths of canyons, in tunnels that burrow through the hearts of mountains, in a canal that connects the Atlantic and Pacific Oceans, and in vast railway systems in various countries of the world.[1]

The world owes a debt of gratitude to him and those like him: pioneers, trailblazers, engineers, builders, and others who conquered nature for people's comfort and convenience.

John Stevens entered the world on April 25, 1853, at West Gardiner, Maine. He was the son of John Smith Stevens and Harriett Leslie French. His ancestry traces back to

---

[1]Marion T. Colley, "Stevens Has Blasted and Bridged His Way Across America," *The American Magazine* (February 1926): 16.

Henry Stevens, who, with his wife, Alice, came to the American colonies from Cambridge, England, in 1635.

John grew up on his father's small farm in Maine. He was full of energy, and strenuous work hardened his body and laid the foundation for the excellent health that was to be his good fortune. He did not distinguish himself in his overall studies, but was a good student in mathematics. After elementary school education, he enrolled in courses at the Maine Normal School in Farmington. He became a teacher, but after one year in the classroom, he realized that profession was not to his liking.[2]

One event of his teaching year produced the formation of a class in astronomy. He had taken a walk one Friday night to get some exercise and forget the activities of a hectic day. He came upon a group of older students. Some were not much younger than he, and a number hoped to become seamen. They were star-gazing and inquired if he knew anything about the planets.

"Wouldn't it be great if we could have a class in astronomy," one of them suggested. "It would be a big help when we go to sea."

"I don't see any reason why we couldn't," replied Stevens. "We'll do it." He returned to his lodging and looked up some textbooks on the subject. After studying them, he added astronomy to the school's curriculum.[3]

Following his year of school teaching, Stevens decided to give it up and go into engineering. Years later, when a reporter asked why he chose engineering for a career, he answered,

> I don't know, unless it was the fact that I was born with the wanderlust. I knew that an engineer's life would offer variety and

[2]Ibid., 75.   [3]Ibid., 74.

change. Of course, I had no technical training. It did not take me long, however, to discover that becoming an engineer involved more than the mere announcement of that determination. I set to work at once to acquire the necessary training. But this was not easy to do. It took me many long years, and during those years, like numberless other boys who have gone through the same grind, I suffered black discouragement.[4]

His first engineering position was with a field crew making surveys in Lewiston, Maine. His work consisted of surveying lots and driving wooden stakes in the mud. Following a year of that, he became restless and felt lured to the West by prevalent propaganda that the West was the place for a young man to earn his fortune.

In 1873, at the age of twenty-one, he went to Minneapolis, Minnesota, which was then a town of some fifteen thousand, on the edge of unsettled prairies. His uncle, Jesse Stevens, was an engineer there, and perhaps influenced the city council to hire his nephew. John Stevens became a rodman on the staff of the city engineers, carrying the surveying instruments around the streets. As an engineer's assistant, he held the leveling rod when it was used. Eventually, he received a promise of instrument work.[5]

His roommate in Minneapolis was George W. Knowlton, who later described some of their activities:

He [Stevens] was at work every night while the rest of us were playing. Coming home late, I would find him still at it—pouring over those dry textbooks, filling sheets of paper with figures and plans of things that I knew nothing about. I've seen him sit for hours with his feet cocked, smoking and thinking. He wouldn't

---

[4]Ibid.

[5]*Congressional Record*, House, 84th Cong., 2d sess., 1956, 102, pt. 7, 9285–89, Congressman Daniel J. Flood giving a eulogy on "John F. Stevens, Basic Architect of the Panama Canal."

speak a word, so thoroughly absorbed he would be in some problem or other.[6]

When Knowlton's remarks were repeated to Stevens in
later years, he said, "That was about the size of it. I've never
believed very much in luck. Work was the only open
sesame that I ever had any faith in."[7]

In spite of keeping his nose to the grindstone, he didn't
deny himself any of the gaieties of Minneapolis. He fell in
love with and courted Harriet T. O'Brien of Dallas. However, he was in no position to marry her since he was earn-
ing barely enough to pay his own board. Yet, being in love
helped make him more determined to get ahead.

Railroad construction in North America was starting to
undergo sudden growth and rapid expansion. In Texas plans
were being made to extend a line across that state. John
Stevens believed the project offered a good chance for a
young engineer, so in 1876 he left Minneapolis and went
to Texas. At first he worked without pay, doing surveys on
the Sabine Pass & Northwestern Railway, but after eighteen months the company went broke, and so did John
Stevens. His meager savings of four years had gone into
living expenses—now he was penniless with no prospects
of a job. Married to Harriet O'Brien on January 6, 1877, he
now had a family for which to provide. Reminiscing about
that situation in later years, he remarked,

> The period which followed was the most difficult I ever experi-
> enced. I found it almost impossible to get any sort of work that
> would enable us to live. I was not at all finicky about it. I did any-
> thing I could find to do. For months I worked on the section—
> laid crossties, and drove spikes, at a dollar and ten cents a day.
> Then I was foreman, and after that roadmaster.[8]

---

[6]Colley, "Stevens Has Blasted and Bridged His Way Across America," 76.
[7]Ibid.   [8]Ibid.

John and Harriet O'Brien Stevens at the time of their marriage.

It seemed to him he had struck rock bottom, and sometimes he must have wondered if people who just drifted along with the current and let the future take care of itself were not wiser than he. But he wouldn't submit to that kind of attitude. As he once told a reporter, "My philosophy is a conviction that if I were ever to escape from drudgery it would be through preparing myself for something better."[9] He stayed and worked in that situation, and before long things did get better.

In 1879 he heard that the Denver & Rio Grande Railroad was planning some extensions to its line. He applied

---

[9]Ibid., p. 78.

for a job and was hired as an assistant engineer. His work was to help locate and construct a railway through the mountains of Colorado and New Mexico. He moved in 1879 to southern Colorado and later to New Mexico. There he established an enviable reputation for his planning of the extremely difficult route through steep canyons of the Denver & Rio Grande narrow-gauge railroad. At last he began to do civil engineering as he had conceived it to be when he adopted it for his life's vocation. It had taken seven rough years to get a start, but from this time on he was never unable to find employment.

Civil engineering work on railroads was difficult to learn. There were few, if any, formal schools teaching it. To become a location engineer, one had to begin as a surveyor and advance toward that goal. Stevens learned to survey preliminary lines between two points and select the best one for the route on which the steel tracks would be laid. When considering distance, a light grade was more important. Engineers working on route location often consisted of a team with a chief engineer, transit man, front chainman, rear chainman, rear flagman, stake marker, stake driver, level man, rod man, topographer, and assistant draftsman. Sometimes their work was very dangerous and strenuous. Long before track-laying gangs appeared in the construction of a new railroad, a small party of civil engineers would have gone through that route with surveying instruments. They left no trace of their going other than a few stakes driven here and there. They were the men who determined where the tracks should be laid.

He remained in New Mexico for nearly two years, and in 1880 moved to Iowa, where he worked another two years as an assistant engineer for location and construction on the Chicago, Milwaukee & St. Paul Railway. The time he spent on that project enhanced his professional competence. He learned fast—first through his own mistakes and then by observing the performance of skilled engineers who had to cope with the area's natural drainage system. Many streams flowed over wide valleys in this area. These had to be crossed at right angles, and the project required expert planning to secure the best location for the rail route.[10]

---

[10]Ralph Budd, "Address at John Fritz Medal Presentation to John F. Stevens," *Civil Engineering*, New York, March 23, 1925, 18–30.

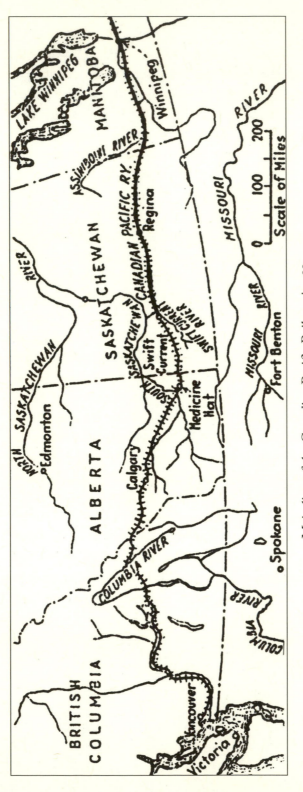

Main line of the Canadian Pacific Railway in 1882.

The main line of the Canadian Pacific Railway in the spring of 1882 was built to a point about 180 miles west of Winnipeg. By the end of the working season it had reached the Swift Current River.

2

RAILS ACROSS CANADA

In 1880, Prime Minister Sir John A. Macdonald met with a group of Montreal businessmen and formed the Canadian Pacific Syndicate. This group was to be responsible for financing and constructing a railroad from Winnipeg, in present-day Manitoba, to the Pacific coast. Members of the syndicate were George Stephen, president of the Bank of Montreal; his cousin, Donald Smith, chief commissioner of the Hudson's Bay Company; Richard B. Angus, a banker; and James J. Hill, a railway promoter and financier.

Building a rail line to the Pacific was one of the largest railway projects ever undertaken at that time. Millions of acres of wilderness had to be surveyed and mapped in order to find a route through the Canadian Shield and mountain ranges of British Columbia. It would be a feat of construction that would amaze the world.[1]

John F. Stevens had finished his work on the Chicago, Milwaukee, & St. Paul in 1882, and was considering other employment. He was challenged by the enormity and vital importance of constructing the Canadian rail line, named The Canadian Pacific. He applied for a position and was hired as a contractors' engineer. His decision to help build

---

[1]David J. Mitchell, *All Aboard! The Canadian Rockies by Train* (Vancouver, B.C.: Douglas and McIntyre, 1995), 47–49.

that line placed him, along with others, in an important role in the development of a new nation.

On July 1, 1867, Canada had become the first federal union in the British empire. It consisted of four provinces: Ontario, Quebec, New Brunswick, and Nova Scotia. It extended westward to the eastern boundary of what would become the province of British Columbia. Sir John A. Macdonald was elected the first prime minister and served two separate terms. When British Columbia was urged to become part of the organized dominion, it agreed on one certain condition—that the new government build a railroad from the Pacific Ocean to the rest of the nation. Macdonald consented to the bargain, and British Columbia became part of the new Canada in 1871. Since the government was financially unable to fund the construction of the railway, the prime minister arranged for private business to build and operate it.[2]

A difference of opinion arose over which route the railroad should take going west out of Winnipeg. Sandford Fleming, a leading Canadian authority on railway surveying and engineering, argued it should be over the northern Yellowhead Pass. James Hill insisted it should be a shorter southern route. However, no one was certain that an acceptable pass through the Rocky Mountains could be found. Hill hired an American railway survey engineer, Major Albert B. Rogers, to search for a more southerly pass through the Selkirk and Rocky Mountains. Rogers explored those western ranges for a year and a half. He finally determined that the Kicking Horse Pass would be a viable route through the Rockies to the Selkirks; the latter range could then be traversed over a pass that he had discovered and which now bears his name, Rogers Pass.

----

[2]Ibid., 30–31.

When the western railway route had been settled, James Hill then recruited another American, William C. Van Horne, to be general manager. He had managed midwestern American railways and was a man of excellent organizing abilities. Van Horne assumed the job of overseeing the entire project in January 1882.[3]

John Stevens went to Winnipeg in 1882 to begin his new job. The contractors kept at least 125 miles of work constantly in progress. As their engineer, he had to go back and forth to observe the subcontractors. He checked their bid estimates to see if they were getting full credit for the amount of work they were doing, and to make sure they were not overpaid by head contractors. Most of his traveling was at night, during the five hours of darkness in that latitude in the middle of summer. After working all day with the contractors, he would drive most of the night in a buckboard to determine how the rail line was to go. He would often turn the reins over to the boy who accompanied him and get in his quota of sleep.[4]

His salary was low, but he learned a great deal about selecting routes for a railway and how to lay tracks properly and efficiently. That knowledge became of great value to him in later years.

The main contractors that first year were R. B. Langdon of Minneapolis and D. C. Shepard of St. Paul. They were experienced and productive builders. Every evening the rail officials could depend on a certain number of miles of track having been laid that day. Shortly after the last daily rail was spiked and the telegraph line alongside it was operating, the crew would gather around to get news from the outside.

---

[3]Ibid., 49–51.

[4]John F. Stevens, "An Engineer's Recollections," *Engineering News-Record*, 1935, 1–70. Cited hereafter as "An Engineer's Recollections."

The year 1882 was a short one for construction. Snowy weather prevented work from beginning before May 15 or continuing after December 1. Yet between those dates, working with great speed, the men laid one thousand miles of track ready for operation as far as Swift Current River. They had also nearly completed the grading for the rest of the line to the South Saskatchewan River. When work was terminated, Stevens returned home to Minneapolis to spend the winter.[5]

In March 1883, Canadian Pacific hired him to be an assistant engineer of location in the mountains. He was assigned to work under Major Albert B. Rogers. Leaving Minneapolis, he went to Winnipeg and then rode over the new tracks to some forty miles east of the South Saskatchewan River where the rail line ended, about forty miles east of the present town of Medicine Hat, Alberta.

Thirty carts with supplies and camping equipment were parked there. Enough engineers and workers arrived to be divided into three full location crews, each assigned to work in a certain area. Stevens' crew pushed their carts as far as they could be taken to a place called The Gap, now the town of Canmore, Alberta. It was the entrance to the mountains located some forty-five miles west of the present city of Calgary. After resting there, they went to what was then the new village of Calgary for supplies. It consisted of about twelve log houses and a main building that was a trading post that supplied trappers, miners, and a few cattlemen. The village was supplied by freighter outfits known as "bull teams." They hauled equipment from Fort Benton, Montana. The day Stevens' crew arrived, some bull teams had just come in. Among the supplies they brought were a half-

[5]Ibid., 2–3.

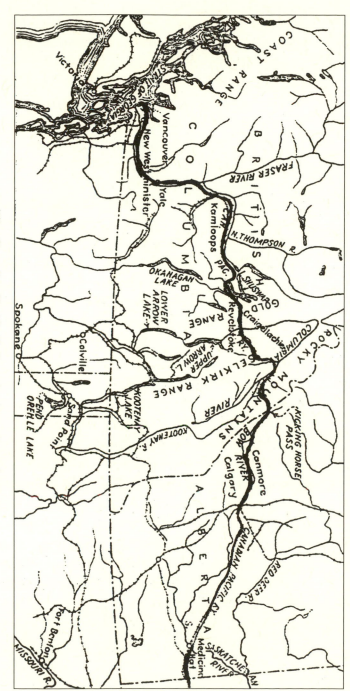

The Canadian Pacific Railway in the spring of 1883 ended about forty miles east of the South Saskatchewan River. Transportation west of there was by Red River carts as far as the entrance to the mountains, about twenty-five miles west of Calgary, and by pack animals through the mountains.

dozen boxes of apples. They were the first ones seen in that part of the West. As a token of hospitality, the freighters presented Stevens with two apples. He was especially grateful, as the others were being sold for fifty cents apiece.[6]

He and his crew remained a few days in Calgary and returned to The Gap to obtain more horses to supplement the few they had. They bought thirty from the Stony Indian Tribe, but got them by having to pay an Indian guide to accompany each pony. The crew left The Gap and began the extremely difficult trip over the Rocky Mountains. They headed northwest up the Bow River and turned west to cross the Rockies through Kicking Horse Pass.

The pass got its name from an event that had happened twenty-five years prior. In 1858, the British government commissioned an expedition to many various areas of western Canada. Dr. James Hector was the leader. One morning they were following along a river bank when one of the horses' packs came loose. The horse lost its balance and tumbled into the river. The water was deep and the bank steep. The men all dismounted and rushed to save the floundering horse, which carried their food supply in its pack. After a struggle, they were able to get the horse on solid ground. Dr. Hector returned to his horse, which was browsing with its lines trailing. When he reached for the lines, the horse whirled and kicked him with both feet in the chest. Dr. Hector was knocked unconscious for over two hours, but in spite of severe pain, he continued with the expedition.[7]

In laying out the railway route, the crew used packhorses to carry the instruments and supplies wherever possible.

---

[6]Ibid., 5.
[7]Mitchell, *All Aboard! The Canadian Rockies by Train*, 100.

When conditions made this impossible, they strapped to their backs the packs containing provisions for two or three weeks, as well as instruments, tents, blankets, and other necessities, and hiked many miles farther into the wilderness. While exploring the mountains, they would have to skirt the dizzying edges of precipices thousands of feet deep, where a single misstep would have brought death. When staking out the course for drillers and blasters, they had to tie ropes around their waists, secure them to trees, and hang over the cliffs. They not only had to blaze their own trail, but also had to build wagon roads to be used by the workers who constructed the line later on.[8] The crew reached the summit of Kicking Horse Pass and had to shovel their way through seven feet of snow.[9] Then they made their way down the Kicking Horse River to the Columbia River, where they set up a camp in the area now occupied by the town of Golden.

After they arrived at the Columbia River, the crew sent their Indian guides home and began work in a summer that was pleasant and mild. Major Rogers, Stevens' supervisor, came twice to inspect the crew's work. He stayed just a couple of days each time and left the work of laying out the railroad's location to Stevens' judgment. Rogers was a monomaniac on the subject of food. He believed that a variety and large quantities of it were not conducive to physical and mental fitness. Therefore, the survey crew's menu consisted of beans and bacon three times a day! Their line of supply in that location began at Spokane, Washington. It went by rail to Sandpoint, Idaho, then by packhorses through the wilderness, to the head of the Columbia River, and finally on to Stevens' camp by canoes.

---

[8]Colley, "Stevens Has Blasted and Bridged His Way Across America," 8.
[9]"An Engineer's Recollections," 7.

Near present-day Golden, the rail line crossed the young, north-flowing Columbia and went west into the Selkirk Mountain Range. With autumn setting in, cold weather arrived. Trails became coated with ice, and the crew lost some of their horses when the animals slipped on the treacherous edge of canyons and were dashed to death down the rocky mountainside.

Falling snow began to make work impossible for Stevens' crew. The other two surveying crews had been surveying the wilderness a few miles east of Stevens' group, and they received word to return home from the mountains. Stevens' party was ordered to remain in the Selkirk area and make a survey over another route, which a trapper had persuaded the railroad company to believe was better than the one they had worked on all summer. Stevens ordered more supplies for the crew and began surveying the suggested new course. After a month of work, they found that the alleged route was of no value.

During their added time of searching, they had not received any of their supplies nor heard from anyone. Taking six men and fifteen horses to carry their equipment, Stevens set forth to find a way out of the mountains. They back-trailed east toward Kicking Horse Pass. The trail became impassible, and in one hour they lost all the animals when they slipped off the icy trail into the canyon hundreds of feet below.[10]

The small party then returned to the other members of the crew, and they proceeded with difficulty to their old camp on the Columbia River. Supplies still had not arrived. Stevens remembered another pass through the mountains at the head of a creek he had last surveyed. He told his men

---

[10]"An Engineer's Recollections," 8.

he was going to try to hike out that way and get help. He asked for volunteers to go with him, and only one man responded. They took a little food, blankets, and Stevens' notes of the summer's work and started east on an unknown creek, which Stevens had been told sometime earlier led by way of a pass to a tributary of the Saskatchewan River east of the mountains. They crossed the pass with difficulty, and suddenly overtook a party of engineers who had been ordered out earlier. They traveled with them for several days but Stevens felt the progress was too slow. He was anxious to get help for his crew left behind, so he and his companion departed to take a route of their own. They took with them a small amount of the engineers' food. They went two days with only one slight meal, snowshoeing with weather much below zero.

Stevens' traveling companion was large and powerful. On one of the days, perhaps because of fear or hunger, he began talking to himself, muttering several times something about cannibalism. Stevens didn't like that, and since his friend was carrying the axe, he thought it would be best to keep the man ahead of him. That made it necessary for him to break trail, which Stevens had been doing most of the time.

Suddenly they stopped and listened. "I hear a dog," Stevens said. "No," said the other, "'tis a tree squeaking in the frost you are hearing. There ain't a dog within a million miles." Stevens pushed on as fast as he could, and all at once a large dog bounded through the snow barking to greet him! Right behind came a search party, and the two weary travelers got a real meal, the first in two days.[11]

The relief group was well equipped with large supplies of

[11]Ibid., 8–9.

food that the railway company was sending to Stevens' crew. Because they had failed, as Stevens' crew had done, to negotiate a passage by way of Kicking Horse River, they were late arriving, having had to detour to the route where they met Stevens and his companion. They went on to the camp on the Columbia River following the trail Stevens and his helper had made coming out. They brought back all the workers to the east of the mountains in good condition. Stevens hiked to the new railroad tracks that had been laid near the Bow River a few miles east of the summit.

He rested there a couple days and then went by rail to Winnipeg and on home to Minneapolis. He found that his wife and friends had about given him up because the newspapers had reported him hopelessly lost, until the news came of his return across the mountains.

Rail tracks for the Canadian Pacific Railway were laid within five miles of Kicking Horse Pass by the end of 1883. During that year the company lost one of its most important developers. While a difficult route was being built on the eastern portion of the rail line north of Lake Superior, James Hill vigorously proclaimed that the eastern portion was an expensive mistake. He believed there was more advantage in having the Canadian Pacific line follow a route south of the Great Lakes through American territory and linking up with United States railroads. However, the government insisted on an all-Canadian route. In frustration Hill resigned from the Canadian Pacific in May 1883. He returned to the United States and eventually built his own railway empire.[12]

---

[12]Mitchell, *All Aboard! The Canadian Rockies by Train*, 54.

*　　*　　*

John Stevens spent the winter of 1883–84 idly at his home in Minneapolis. The Canadian Pacific Company again sought his services, and he contracted to work another year with his former boss, Major Rogers. This time they were to take charge of the location of the railroad's main line, continuing west from the Selkirks across the Columbia River and through the Gold Mountain Range (now the Monashee Range) in the southeast part of British Columbia. The crossing was some three to four hundred miles inland where the river flows a southerly course between the Selkirk and Monashee ranges, close to the present town of Revelstoke, British Columbia.

In March 1884, Stevens left home and went with a group of about twenty-five men, including two other engineers, to San Francisco. Before getting out of Minnesota, they were caught in a snow slide, and were blocked again by snow in the Sierra Nevada in California. It took some twelve days to arrive in San Francisco, and they missed connections with the ship that was to take them to their destination in Canada. Following the delay they boarded another vessel and reached Victoria, British Columbia, after a rough three-day voyage. They went across the Strait of Georgia to New Westminster, just south of Vancouver, and continued by steamboat up the Fraser River to the head of navigation at Yale, British Columbia. In Victoria and New Westminster they picked up enough workers to compose three full locating crews.

The rail line going east was constructed from Vancouver as far as Yale, so the crews had to travel on by foot to the village of Kamloops. Major Rogers met them there and assigned the crews their areas of work. Stevens' crew had to travel by steamboat up the South Thompson River to the head of Shuswap Lake. From there it was all wilderness. Rogers had promised them a train of packhorses and packers, but either through negligence or some whim at headquarters in Kamloops, the horses did not arrive.

Stevens' men not only had to cut and build miles of trail through the rough country across the mountains to the Columbia, but they also had to pack their tents, cooking equipment, baggage, and a month's food supply on their backs. By the time the promised animals reached the party, it was the middle of the summer. Stevens was understandably provoked that engineers and their helpers had been obliged to make packhorses of themselves. It was a monstrous struggle with a pack strap across the forehead, toiling up a muddy trail, carrying a pack made up of a grindstone and a sack of flour, all the while trying to absorb knowledge of the topography through which they were passing.

After floundering through deep snow, climbing over immense logs, and wading through swamps that they had to corduroy before they could cross, they made their camp on the summit of the Monashee Mountain Range, seven miles from the Columbia. The weather was bad that summer, and because of the frequent rain the workers were seldom dry. Regardless, they earnestly worked at their surveying and hurried their task as fast as possible through the thick forests and along the mountain sides.[13]

As the Canadian Pacific line progressed in the western section where Stevens and his crew worked, railroad contractors in the eastern area confronted an enormous problem. When they reached the summit of Kicking Horse Pass, they encountered a staggering drop. There was no room to lengthen the line or reduce the gradient. The Canadian

---

[13]"An Engineer's Recollections," 11–12.

Pacific Company obtained government approval for a steep grade of 4.5 percent on the western slope of the pass, which was safe for a rail line. The location became known as the Big Hill and was permitted as a temporary measure so that construction could continue. The rails were laid in an eight-mile-long breakneck descent from the top of the pass to the present town of Field, dropping 237.5 feet to the mile. It was considered the most hazardous stretch of railroad in Canada. Surprisingly, not a single accident involved a passenger train in the first twenty-five years that trains ran on the Big Hill, although laborers lost their lives on freight and work trains.

Traffic increased to the amount that the railroad realized they had to do something about the temporary steep grade. Their solution was to construct spiral tunnels, a concept that had been used previously in Europe. Two complete circles were tunneled into the steep mountains on both sides of the Kicking Horse River, with the tracks crossing over themselves in a figure-8 pattern. This increased the length of the main line by almost five miles and reduced the grade by more than half. The project was considered as one of the engineering marvels of the world.[14]

Late in the season of 1884, the provincial government of British Columbia built a wagon road through the Monashee Range following much of the pack trail that Stevens' crew had built. About the same time, the Canadian Pacific signed the contract to build the railway line that Stevens had located through the mountains from the summit east to the Columbia River, and he was given charge as engineer of its construction. That meant he would stay there another year. Acting on her own initiative, his wife, Harriet, decided she

---

[14]Mitchell, *All Aboard! The Canadian Rockies by Train* , 71, 74, 77, 80.

would join him. She and their two-year-old son left their home in Minneapolis and went to Portland, Oregon. They boarded ship there and sailed to Victoria, where John Stevens met them, and together they traveled by steamboat on the Fraser River to Yale. The rail line from Yale had been extended, so they rode in the caboose of a construction train to Kamloops. From that town they took the steamboat up the Thompson River, over Shuswap Lake, to The Landing. The final lap was taken by driving a team on the new, rough wagon road over the range to the Columbia River.

The general contractor, C. B. Wright, had built a large and comfortable log house by the river where he lived with his wife and two grown sons. The Stevens family lived with them during the winter. Mrs. Wright and Mrs. Stevens were the only white women within a hundred miles in any direction. Everyone was congenial, and all enjoyed a comfortable and happy winter.

Work that season involved clearing the right-of-way, cutting cross ties and timber, and installing wooden crib piers for a bridge across the Columbia River, located near present-day Revelstoke. Stevens conceived the idea of making cribs and sinking them in place during the winter. Crews waited until the river froze solidly from shore to shore and the ice was thick enough for teams of horses to carry heavy loads, then made cribs out of squared 12″ × 12″ timbers. Cutting holes of comparable size in the ice, they filled each crib with rock and sank it to the hard gravel on the river bottom. Piers were made and set on the cribs, and the bridge was completed by spring.[15]

When the spring of 1885 came, Stevens, out on the western line, moved his engineers to various sections so they

---

[15]"An Engineer's Recollections," 12–13.

could speed the work of grading. Completion of the Canadian Pacific Railway was getting close. William Van Horne, general manager, went to view their work. En route was Summit Lake, hemmed in by bluffs and which had to be crossed on a raft. As they were crossing, the general manager fell into the icy water. When Stevens pulled him out, he was amused at his boss's vigorous and breezy comments about the country and everybody connected with it.

After four years of hard labor, construction of the Canadian Pacific Railroad was finished. Rail lines from both east and west were designated to join together at a particular place, now known as Craigellachie at Eagle Pass in the Monashee Mountains. It so happened that the location was exactly the site where Stevens had built his crew's summer camp. The western line from Vancouver was completed October 1, 1885, and the eastern line from Winnipeg was finished November 7, 1885.

A simple service at Eagle Pass was held on November 7. Donald Smith, a member of the syndicate that had financed the construction, drove the railway's last spike. His first blow was too feeble, and the iron spike bent. A spare one was quickly set up, and Smith hit it with careful, precise blows, driving it into place. Officials and workers present cheered and called on general manager Van Horne to make a speech. His message was stated in just a few words: "All I can say is that the work has been done well in every way."[16]

Years afterward, during a news interview, John Stevens recalled his Canadian experiences:

> During those years, I became tough and hard physically. I learned to sleep under wet skies on the Western plains, rolled only in a single blanket, and to adapt myself to the rigors of a far North-

---

[16]Mitchell, *All Aboard! The Canadian Rockies by Train*, 61.

ern winter, under the most primitive conditions. There is no other sort of work connected with engineering that compares with locating and surveying. The happiest years of my professional life were those spent with the Canadian Pacific and later, the Great Northern, bridging and blasting highways through the unknown West. It was then I did my best work.

He had been in personal contact with a number of perplexing problems and had managed to solve them. Finding solutions to them gave him a strong sense of confidence which he felt was unshakable.[17]

When Stevens' services were no longer needed, he and Harriet began to think about going home to Minneapolis, and they arranged for Harriet and their son to ride there in the private railroad car of James Ross, a manager of construction for the c.p.r. She and the boy had a comfortable journey and arrived home, both in good health, in December.

Stevens wanted to see something of the Lower Columbia River country, especially Spokane, before leaving the area. He found three other men who also wanted to go south. They bought a small rowboat and started down the Columbia. After several days of rowing and floating, they crossed the boundary line into the United States. From that place they felt the navigation of the river was impractical, so they hired a settler to drive them to Colville, Washington. They went on by bus to Spokane, eighty-five miles farther. After arriving in Spokane, they learned they had been followed all the way by a small gang of thugs who thought members of Stevens' party were carrying a lot of opium that had been stolen up on the Canadian Pacific and were smuggling it into the United States. One of the hoodlums was hanged for a brutal crime he committed shortly afterward.[18]

---

[17]Colley, "Stevens Has Blasted and Bridged His Way Across America," 81–82.
[18]"An Engineer's Recollections," 15.

3

Constructing
American Railroads

Early in the spring of 1886, the chief engineer of the Chicago, Milwaukee & St. Paul Railroad asked John Stevens to take charge of the construction of a ninety-mile rail line in Iowa from Sioux City to Manilla. It was a line he had located when he worked for that railroad in 1882. With the Canadian Pacific project finished and no other employment in sight, he eagerly accepted the job. As he stated in his recollections: "A field engineer in those days was much like a bear; if he hadn't laid on a reasonable amount of fat during the summer season, he had hard work to hibernate through the winter."[1] After his experience in the mountains, the building of this line seemed tame to him. The work went smoothly, despite some problems that challenged his skill as an engineer, and on the whole he felt satisfied with the results of the season's work. With the grading and bridging completed, there was no prospect of more employment with the Milwaukee.

Fortunately, he was offered and accepted a position in northern Michigan. A syndicate had been organized to create a railroad that would extend from Sault Ste. Marie, Michigan, near the International Bridge, westward through

---

[1] "An Engineer's Recollections," 16.

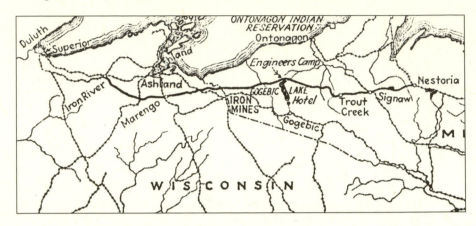

Construction of a through railroad line from Sault Ste. Marie
to Duluth was undertaken in 1886. It called for new work from
the International Bridge at Sault Ste. Marie to Soo Junction
and from Nestoria to Iron River, as indicated in heavy lines.
From Iron River to Duluth the South Shore R.R. operated
over Northern Pacific Railway tracks.

the northern part of the Upper Michigan Peninsula and into
northern Wisconsin to Duluth, Minnesota, on Lake Superior. They owned a line known as the South Shore, which
went west from Soo Junction to Nestoria, Michigan. They
bought two other railroads. One extended from Marquette
to the Straits of Mackinac at St. Ignace. The second ran from
Marquette to the west through the iron range and then
northwest through the copper region to Houghton, Michigan. Those acquisitions then left 40 miles to be built from
Sault Ste. Marie westward to Soo Junction, and about 220
miles west from Nestoria to Iron River. From Iron River to
Duluth the new railroad would run over the Northern Pacific
Railroad tracks. Stevens' salary was $250 per month, twice

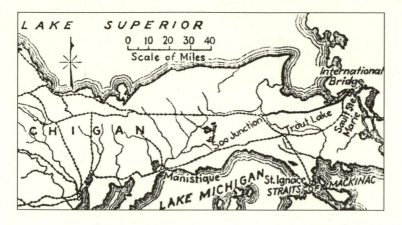

what he had ever received. He was overwhelmed at first with its size, but later realized the responsibilities he had assumed were very great and worthy of the pay. He closed out his work in Iowa and went to Marquette, Michigan, where he established his headquarters for the next two years. His family went with him.[2]

The chief engineer for the South Shore line was J. A. Latcha, whose office was also in Marquette. He had placed several teams in the field to work on location surveys. Latcha was an experienced engineer. He had been chief engineer on the Nickel Plate Railroad during its location and construction. Nickel Plate was a nickname for the New York, Chicago & St. Louis Railroad, which ran from Buffalo to Cleveland, into Indiana, and on to Chicago. It was part of W. H. Vanderbilt's Central Railroad System. The line was given that nickname partly because the employees complained and even quit because the passengers' tipping was so skimpy. Latcha had also held important positions on

---

[2]Ibid.

other lines. Now, he was contented to remain permanently in Marquette and let Stevens, the assistant engineer, attend to the active work connected with locating and constructing the new railroad. He deferred major decisions and all matters connected with them to Stevens' judgment. In fact, he never saw a foot of the rail line or of the country through which it passed until he went over it on a special train a month or more after its completion.[3] Stevens later commented, "I did the whole thing. I made the surveys, located the roadbed, laid the tracks, and put the trains on them."[4] In spite of remaining apart geographically, Latcha was a person with whom Stevens easily could work, and he later recalled that their two-year association had been harmonious in every way.

Stevens began his new work in the month of December. It was a very cold winter accompanied by deep snow. Much of the country through which the lines were projected was a wilderness. For one stretch of one hundred miles, there was not a house, much less anything in the form of a settlement. There was just one road, the old Ontonagon military road that ran north clear across that part of Michigan to the Ontonagon settlement on Lake Superior. It had become nearly impassable due to disuse. Snow that already had fallen averaged at least three feet deep over that section of country. With the short days of winter, the prospect of locating the roadbed during that cold season was not encouraging.

Because of the lack of roads and the swampy nature of the terrain, it was vital for the construction contractors to get their equipment, camps, and large amounts of supplies

---

[3]Ibid., 16–17.
[4]Colley, "Stevens Blasted and Bridged His Way Across America," 82.

in place before the spring thaw. They couldn't do that, however, until they determined the final location of the roadbed, which would indicate where the camps should be placed. It was another winter of extremely strenuous work for Stevens. His first priority was to go over the country through which the railway was to be built and decide where the location should be. He made changes in the plans and methods that were being followed by the surveyors in the field. He also made changes in and additions to the personnel of the engineers.

The problem of keeping workers supplied with food in that wilderness was difficult, since the transport of provisions was done by packing on men's backs or by toboggans. In the case of some of the most isolated camps, it was done by dog teams. There were ten large groups—with about ten to a group—to supply, so Stevens organized and equipped a large convenient commissary. It was very effective, and there was never a shortage of food or other necessities at any camp during the winter.[5]

He spent almost the entire winter on snowshoes traveling from one crew to the next. Sometimes he had a man accompanying him, but mostly he traveled alone. Those in the Marquette office seldom saw him. It wasn't long before he had the engineers' work assigned and the surveys progressing rapidly in spite of the handicaps, one of which was the necessity of digging through several feet of snow to set every stake. Before the snow melted and the swamps thawed so as to become impassable for work teams, the greater part of the final location of the roadbed had been decided. The contractors' camps, equipment, and supplies were now ready for construction of the roadbed to begin.

---

[5]"An Engineer's Recollections," 17.

In areas where deer were fairly plentiful, timber wolves roamed the wilderness. Late one afternoon while travelling alone in a swampy area near Soo Junction, Stevens came upon a group of five wolves. There was only a small amount of snow on the ground, so he slung his snowshoes on his back and ran until he found a tree he could climb. The animals kept him treed for a couple of hours, then just before dark, for some unknown reason, they went away as suddenly as they had appeared. As soon as the wolves were out of sight, Stevens dropped to the ground and scurried in the opposite direction, charting his course to stay within running distance of climbable trees. In brief time he reached the railroad and followed it to the first place that offered him shelter for the night.[6]

On another occasion he came by rail and got off the train about five miles south of Lake Gogebic, which was covered with ice. A camp of engineers was based at the north end of the twelve-mile-long lake, and Stevens wanted to visit it. After staying overnight at a so-called summer resort hotel at the south end of the lake, Stevens began his trek to the camp early the next morning. He had been warned that the lake ice was not safe for travel—much of the surface snow on the lake had been blown away, and a few warm days had caused the ice to melt partially. Cold nights had caused the surface water to refreeze, but the ice still was not thick enough for his travel. Nevertheless, he headed north on the ice, and it held him up for a while. Then he began to break through into six inches of water. Soon, his boot and leg coverings became solid ice. Determinedly, he forced himself to slush through the icy water for the last ten miles to the camp, following close to the lake shore.

---

[6]Ibid., 18.

He reached it just before dark, completely exhausted and nearly frozen. Fortunately, he survived by getting thawed out, eating nourishing food, and having a good night's sleep.[7]

As the summer of 1887 approached, Stevens increased the work forces as fast as laborers, most of them from Chicago, could be recruited. The work of all the contractors was satisfactory except for one firm from New York City, which had neither appropriate equipment nor experience for that type of construction. As it turned out, they didn't have enough capital either, but with the help of the railroad company, they managed to finish their contract without holding up the laying of tracks.

Grading and bridging continued from 1887 into 1888, and finally in the fall of that year, the work was completed. Tracks had been laid and sand ballasted, telegraph line and buildings erected, and final touches added. Toward the end of construction, one of Stevens' gangs of Italian workmen took a violent dislike to him, because he had ordered a contractor to burn the carcass of a mule that had died from natural causes. The gang wanted to eat it. When an opportunity arose, several of them chased him a half-mile down the tracks, flourishing knives and clubs accompanied with wild shouts. Stevens was too fast for them, already having outrun wolves, and was able to reach safety easily. Other employees thought it was a good joke on the "old man" for presuming to restrict the "dietary" of his workers.[8]

As an engineer, Stevens took little pride in building that particular railroad. He had been advised when he took charge that the principle reason for constructing it was to make money for some members of the syndicate that had financed

---

[7]Ibid., 18.   [8]Ibid., 19.

it. After it was completed the chief engineer told him that he had enabled a profit to be made of about a million dollars more than had been estimated before he took charge of its location and construction. Though Stevens was properly paid his salary and thanked for his excellent efforts, he never saw any of the extra million-dollars profits.

The new railway was finished in 1889, but for two months it looked as though the road would have to remain idle or operate itself, because the syndicate could not find a buyer for the South Shore line. Finally, the Canadian Pacific Railway, which connected with it over the International Bridge at Sault Ste. Marie, took it over. William Van Horne, the C.P.R. president, and other officials made a trip over the new road by special train. It was the first passenger train to travel over the entire main line from Sault Ste. Marie, Michigan, westward to Duluth, Minnesota. J. A. Latcha, chief engineer under whom Stevens served, was on the special train. At that time, he got his first look at any part of the finished road.[9] The new rail line became known as the Duluth, South Shore & Atlantic Railroad.

When the new road went into operation, Stevens was asked to stay and be the superintendent, but he wanted to remain with engineering and refused the offer. He ended his connection with the company and went home to Minneapolis. However, his association with that railway was not completely closed. The firm of contractors from New York, which had not done very good work, refused to make its final settlement with the company on the basis of Stevens' cost estimates. They sued for a half-million dollars more than the estimates allowed them. One part of their claim charged him with "general oppression"—which meant that

---

[9]Ibid.

he had kept them close to the specifications, including classification of materials handled in executing their contracts. The trial was held in Detroit, and he attended for six weeks as principal witness for the company. The verdict was in favor of the contractors for the amount of $60,000, which was a small amount as it included $50,000 of retained percentage—money withheld by the government to pay contractors and to ensure that contractors had fulfilled their obligation—which they could have had at any time, without the inconvenience of a trial, if they had made a final settlement.[10]

While Stevens was working on the Duluth, South Shore, & Atlantic Railroad, the Russian government sent two of its engineers to study the American system of railroads. He met them several times and was able to offer them small courtesies. The next year the Russian government invited him to visit Russia and inspect its railways. They would pay his expenses. He was tempted to do it but was unable. Little did he know then that thirty years in the future he would spend more than five years in that country during World War I.

America's Northwest again beckoned to Stevens. Its towering mountains and unsettled wilderness held vital challenges for him. In the spring of 1889, he went to Spokane, Washington, to seek employment. Living there was Spokane's railroad builder, Daniel Chase Corbin, who eventually built six railroads in that area. He sold some of them to the Northern Pacific Railroad, which later sold them to the Great Northern Railway. Stevens arrived in Spokane at the time Corbin was planning to build a line from there to the Canadian border, a distance of 121 miles, and had hired Edward J. Roberts to be the chief engineer.

---

[10]Ibid.

The railroad was to be known as the Spokane Falls and Northern. Roberts met Stevens and hired him to survey the route for the new line. During the summer of 1889, Stevens worked in making the final surveys.[11]

At the conclusion of that job in the fall, Stevens was in Spokane when he received a telegram from Elbridge Beckler, chief engineer of the Montana Central Railroad, a 97-mile-long line that extended south of Great Falls, through Helena, to Butte. It was the property of James Hill's company and was headquartered in Helena. Beckler asked Stevens to meet him in Helena as soon as possible.

Eleven years earlier, James J. Hill in Minneapolis had desired to become owner of the St. Paul & Pacific Railroad. When that line went bankrupt, he and his associates, Norman Kittson, St. Paul agent for the Hudson's Bay Company; Donald A. Smith, governor of the Hudson's Bay Company; and George Stephen, president of a bank in Montreal, bought it in 1878. They extended the line westward through Dakota Territory and then north to Pembina, now in North Dakota, at the Canadian border. The Canadian Pacific Railway ran a line down from Winnipeg, and the two roads met at Pembina in 1879. Hill's group named the newly-organized railway St. Paul, Minneapolis, & Manitoba.

Hill had a strong desire to build a transcontinental railroad. He told his friends he was going to build his new railway across the continent to reach Puget Sound. By 1883, Hill had full control of the properties owned by him and associates. By 1887, he had extended the line from Min-

---

[11]George F. Brimlow, "Marias Pass Explorer," *The Montana Magazine of History* 3 (Summer 1953): 39–44. Mr. Brimlow used information from a letter he discovered written by John F. Stevens to General William C. Brown in May 1928. John Fahey, *Inland Empire—D.C. Corbin and Spokane* (Seattle: University of Washington Press, 1965), 93–94 and 236.

neapolis west to Havre, Montana, and that same year had connected it to the Montana Central Railway at Great Falls. Hill then changed the name of his railroad to the Great Northern Railway on February 1, 1890. This did not include the Montana Central line, as it was operated separately.

Stevens went to Helena and met Beckler, who informed him that Hill was still planning to extend his railway from Havre to Puget Sound and had appointed him (Beckler) as chief engineer of the proposed extension. He explained further that the plans were tentative and incomplete. It had been decided that the extension must follow the valley of the Kootenai River in western Montana and northern Idaho. Flowing south from British Columbia into the northwest corner of Montana, the Kootenai makes its turn to the west in the approximate latitude of Havre. The only stumbling block in running a direct line between the two points was a major one—the Rocky Mountains. To overcome this impediment necessitated finding an unknown but rumored-to-exist pass across the Rockies in the vicinity of that parallel. Otherwise the GN would have to choose one of two more southerly passes and then find a way north through the Flathead Valley. From there the route could ultimately intercept and follow the Kootenai River as it cuts its way west between the Cabinet and Purcell mountains. Such a loop route would place the railroad under a permanent handicap in handling through-traffic in competition with other transcontinental railways, because it would give the GN line much more severe gradients and alignment than was desirable, and would be one hundred miles longer. The problems facing Hill: Did a feasible mountain pass exist and could it be found quickly? Time was pressing, and Hill was not one to delay the execution of his plans once he had made them.

*   *   *

Beckler then told Stevens about an old Indian legend. For a long time it had been known to the Blackfeet and Kalispel Indians that there was a gap at one of the heads of the Marias River. The river rises in a number of places in the Rocky Mountains and unites to form the main stream that flows into the Missouri a few miles northeast of Great Falls.

Famed explorer Meriwether Lewis was certain that there must be such a low gap in the mountains at the head of Marias River, and on his way back from the Pacific Coast in 1806 he led a small group in an attempt to find such a gap. They failed to reach the mountains because of an encounter with Indians. Lewis named the river "Maria's" for his cousin, Maria Wood.[12]

In March 1854, Congress had formed the new territory of Washington out of the northern half of what was Oregon Territory. The boundaries of the new territory extended six hundred miles east from the Pacific Ocean to the crest of the Rocky Mountains in what is now the state of Montana. Except for a few missionaries and trading posts in the interior, the whole vast region was unsettled, and much of it unexplored by civilized men. It contained thousands of Indians, some of them hostile. Congress also appropriated funds for the exploration and survey of four railroad routes from the Mississippi River to the Pacific.

The government's bid for governor of Washington Territory had attached, ex officio, the superintendency of Indian Affairs and also the charge of the exploration of the northern rail routes over the mountains. Army officer Major Isaac I. Stevens (no relation to John Frank Stevens)

---

[12]Brimlow, "Marias Pass Explorer John F. Stevens," 41, footnote 12.

applied for the governship. President Pierce sent Stevens' name to the Senate, who confirmed him as governor. Stevens then resigned his commission from the Army.

The new governor was charged with examining the passes in the Rocky Mountains during the winter when the snow was the deepest, and at early summer when the streams were highest. From 1854 to 1855, he explored mountain passes from the vicinity that is now Havre, Montana, to as far west as Flathead Lake and beyond, but was not satisfied with any he saw. On one of his trips he and his men came upon a group of Piegan Blackfeet Indians. Their chief, Little Dog, told Stevens about a low gap in that region. However, the white explorers crossed several miles north of the gap and didn't find the Marias Pass that they were seeking.

Isaac Stevens established an operation base at Fort Benton and planned to take carts and a few wagons over the mountains west of there. If successful, it would show that route to be suitable for railway use. He assigned Frederick W. Lander, a civil engineer, to lead the exploration and also search for the elusive Marias Pass. Isaac Stevens was going part way with them to the Blackfoot Indian camp to get guides for his survey work. The party left Fort Benton on September 9, 1853, but was called back because an officer had sent them word that it would be impossible to take wagons across the mountains in time to avoid snow storms. They returned to Fort Benton, where Stevens received a dispatch from Secretary of War Jefferson Davis, stating that government funds for the railroad survey project had run out, and he was to cancel the entire program. Isaac Stevens believed the search for Marias Pass was too important to discontinue. He wrote to Davis requesting the exploration be extended for another year. His request was refused. However, Stevens continued the project hoping the government would understand why his venture was so important.

He left Fort Benton and journeyed to the Bitterroot Valley to establish another base of operation. He located his camp near the junction of the Bitterroot and Hell Gate (now Clark Fork) rivers, northeast of the present city of Missoula. He planned to go west to Walla Walla, and he assigned to leaders of his various parties the remaining survey work in Montana. As a final attempt to locate Marias Pass, he sent A. W. Tinkham, a civil engineer, with a small party to continue the search. They traveled to the mouth of the Jocko River, then turned north and followed the lower section of the Flathead River to Flathead Lake. Their trail went around the western shore of the lake and continued north to the upper section of the Flathead River, which flowed south into the lake. Twenty-five miles northeast of the lake the Flathead River forked, one branch coming from the north and one from the east. Tinkham chose the east branch, now known as the Middle Fork of Flathead River. Unfortunately, he left the river too soon and crossed over a mountain pass, now called Pitamakan and then called Cut Bank Pass or False Marias Pass. It was rocky, steep, narrow, and high, and Tinkham found it to be entirely too impractical for wagons to travel over it. He descended on the east side of the mountain to Cut Bank Creek, followed it to the Marias River, and that river to Fort Benton. Isaac Stevens had left James Doty in charge of the work there, and Tinkham reported to him that the pass was of no use for railroad, wagon, or even a two-wheeled cart. The engineer then left Montana and went to Washington Territory to join Governor Stevens, with Marias Pass still undiscovered.

The following May, in 1854, another crew under the leadership of James Doty explored the northeastern slope of the mountains now known as Waterton and Chief. While fording the Two Medicine Creek of the Marias River, they met an Indian who pointed out the location of Marias Pass. He also showed them the pass twenty miles north which Tinkham had crossed. Returning from his trip north, Doty rode to a location where he got a good look at the course to Marias Pass and realized that Tinkham had made a mistake. However, Tinkham was Doty's superior in the survey hierarchy, so Doty never said what he thought. Although the Doty party came within sight of what they believed was the gap of Marias Pass, they didn't take the time to explore further to verify it.[13]

*   *   *

After Elbridge Beckler had related to John F. Stevens the legend and history of Marias Pass searches, he asked the young engineer if he would accept the job of ascertaining if such a route as Marias Pass existed. If so, would it be feasible for a railway to use it as a route through the Rockies? It was November 1889, and winter was starting to set in, yet Stevens accepted the challenge.[14]

Furnished with a letter of introduction, Stevens left immediately and went to Fort Assiniboine, a military post seven miles west of Havre. In his search for Marias Pass, his plan was to make a thorough reconnaissance of the mountain areas. The commandant at the fort gave him full cooperation and equipped him with supplies and horses. He followed the Missouri River south to the vicinity of Great

---

[13]Budd, "Address at John Fritz Medal Presentation to John F. Stevens," *Civil Engineering*, 22–23.

[14]"An Engineer's Recollections," 21–22.

Falls and explored territory north of Butte and Helena. He searched the ranges and canyons in a one-hundred–mile area, but without success. He figured to try next the territory west from Fort Assiniboine.

He returned to that post in late November. Then, equipped with a wagon and team of mules, a wagon-driver, saddle horse, and a man from St. Paul, Stevens continued his reconnaissance. They encountered a constant succession of blizzards, and the companion from St. Paul kept fortifying himself from the cold weather with strong drink, making him useless to Stevens. Traveling about one hundred miles west from Fort Assiniboine, the small party reached the Blackfoot Indian Agency at the present town of Browning. Stevens sent his traveling assistant back, and tried to get some of the Blackfoot Indians to be guides. They still maintained the tribal superstitions that a visit to that area meant death from bad spirits, and none were willing to be hired as guides.

Finally, Stevens came upon a Kalispel Indian named Coonsah. He reportedly had killed a member of his clan and was fleeing through one of the southern passes. The Blackfoot tribe had given him sanctuary. By offering the Kalispel a good wage, Stevens hired him as a guide. The engineer bought some old snowshoe frames from the Indians, and patched them with leather strips so that they became usable.[15]

Stevens, Coonsah, and their wagon driver pushed on from the Blackfoot Agency. It was now December, and the snow was deep. They drove the wagon almost to the foot of the mountains, sleeping at night in abandoned Indian cabins. When deep snow prevented the mules from going farther,

---

[15] Albro Martin, *James J. Hill and The Opening of the Northwest* (New York: Oxford University Press, 1976), 382; and "An Engineer's Recollections," 24.

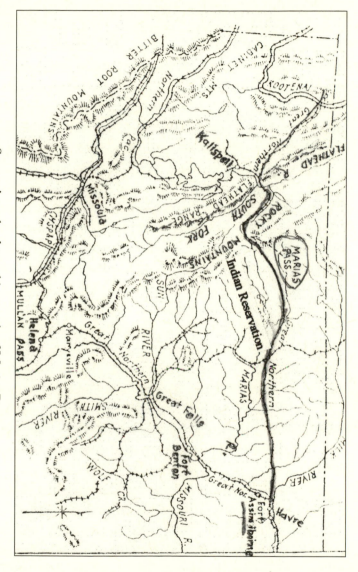

Stevens' overland travel in search of Marias Pass, beginning at Fort Assinniboine and going west.

Stevens and Coonsah continued on by foot, carrying their food and blankets. The driver took the mule team back to the last cabin where they had stayed, to remain there until Stevens and his guide returned.

Actually, the Kalispel had never traveled that far west before and knew nothing of that territory. He was a useless guide, but Stevens kept him to use as a messenger in case of a disaster. The two struggled through the snow on their improvised snowshoes. They reached a small valley that had the appearance of being part of a mountain pass and followed it until they arrived at a location later known as "False Summit." They proceeded on to a small creek. There, Coonsah indicated by sign language that he was sick and could go no farther. Stevens found some dry wood, scraped off some snow, and made a fire. He left the Indian there and went on alone.

Some sixth sense urged him to keep going in spite of the bitter cold. After walking a few miles, he by chance walked right into the present Marias Pass! He continued on down into the western valley to be certain he had crossed the Continental Divide. He observed that the stream of water under the snow beneath his feet was now flowing westward toward the Pacific Ocean. John Stevens discovered the long-sought Marias Pass on December 11, 1889. It was right where Chief Little Dog had long ago said it was. Its elevation is 5,213 feet.

The day was beginning to end. After an exhausting struggle, he got back to the false summit and made a bivouac for the night. The cold had become so extreme there was no question of lying down. Because of the snow, there was no wood to find, only green brush, making it impossible to build a fire and keep it going. Stevens stamped a path in the snow about a hundred yards long, so he could walk without the snowshoes. To keep warm, he walked back and forth. When enough daylight dawned to make it safe to travel, he started back to the Indian, Coonsah. When he reached him, he found that he had let his fire go out and was nearly frozen. They returned together to the Blackfoot Agency and found that the thermometer had registered thirty-six degrees below zero during the night they were on the mountain. On the summit where they stayed, 1,500 feet higher, it would have been even colder.

During an interview years later, Stevens said, "It was a most interesting night, for tramping there in the darkness, on the summit of that mountain pass, I saw as clearly as if it were already a reality, exactly how the trains of the Great Northern would go sweeping through those mountain fastnesses in the months to come."[16]

At the Blackfoot camp he sent the mule team and driver back to Fort Assiniboine. He went by stage, a two-day trip, to Great Falls and on to Helena. He reported his amazing discovery to chief engineer Beckler, then went to his home in Minneapolis, arriving the day before Christmas.

When the news reached James Hill, he immediately ordered workers to cease survey work in other areas and to begin plotting a roadbed over Marias Pass. In his report at a stockholders meeting in 1890, Hill called the new route "an extremely favorable pass."[17]

The advantages that the Great Northern Railway achieved were numerous: the saving of more than one hundred miles of distance, much less curvature, a much better grade, and the lowest Continental Divide railway pass in

---

[16]Colley, "Stevens Has Blasted and Bridged His Way Across America," 86.

[17]Ralph W. Hidy and Muriel E. Hidy, "John Frank Stevens," *Minnesota History* (Winter 1969): 248.

the United States north of New Mexico. Those advantages caused the Great Northern Railway to be the most economical, from an operating standpoint, of any of the transcontinental railroad lines.

In appreciation of Stevens' work, the Great Northern erected a bronze statue of him on Marias Pass, near the spot where he spent that memorable night. The monument shows Stevens as he appeared in 1889, clad in heavy winter clothing.

Over the years, a controversy developed as to whether or not John Stevens was the first one to discover Marias Pass. After study and investigations, the United States Geographic Board, a group appointed by Congress to make final decisions in matters of naming natural features, announced on April 3, 1933:

> John F. Stevens Canyon, Flathead County, Montana, traversed by the south boundary of Glacier National Park between Marias Pass and Belton, . . . For John F. Stevens, C.E., who on December 11, 1889, located and explored the low pass across the Continental Divide. This pass was known to the Indians but there was no record of a white man having knowledge of its location until Stevens, alone, went to and through it until a western flowing stream convinced him of his success.[18]

That quote, which was a ruling of the U.S. Geographic Board, is conclusive verification that Stevens discovered Marias Pass.

He reacted to that pronouncement with the modest reply:

> Whatever hardships and threatened dangers I may have confronted in my quest were compensated for many times over by the fact that the exploration made possible my contribution with the Great Northern Railway and with Mr. Hill, which I look back

---

[18]"An Engineer's Recollections," 26.

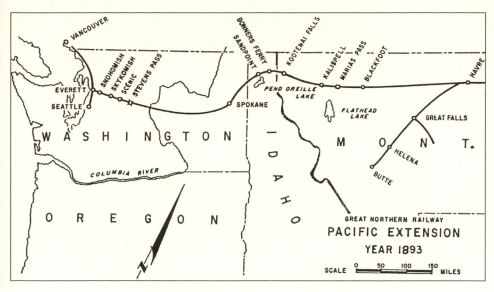

Great Northern Railway's Pacific Extension, 1893.
State Historical Society of Minnesota.

upon as the most satisfactory of my long and diversified career as an engineer.[19]

Stevens remained idle at home during the winter of 1889–90. Early in the spring of 1890, James Hill hired him to take charge of reconnaissance and surveys for the Great Northern Railway's Pacific Extension. He was assigned the task of finding the best route for the railroad from Spokane, Washington, to Puget Sound.

As late as 1894, little information was available about the Cascade Mountains. Until a few years prior, the only known pathways to the Pacific Ocean were the Fraser and

---

[19]Ibid.

Columbia Rivers. Then, in 1888, the Northern Pacific Rail-road built a line over the mountains via the Yakima River and Stampede Pass in Washington. Hill was interested in new country farther north. He gave orders to Stevens to explore every possible crossing between the Northern Pacific tracks and Canada, a territory of 33,000 square miles. Initial thinking was to follow the Wenatchee River west from the Columbia River and over the Cascades to Puget Sound, arriving in the area of the present city of Seattle. Another possibility was to go north from the present town of Quincy, Washington, follow the Moses Coulee across north-central Washington, past Waterville, and down to the Columbia River at Entiat. There the route would fol-low the Columbia north to Chelan Falls, skirt Lake Chelan, cross the Cascades, and descend the west slope of the Cas-cades, reaching Puget Sound at the site of the present city of Bellingham.[20] The major problem in making the deci-sion of the route was deciding where could or should the mountain range be crossed?

Stevens favored a route along the Wenatchee River west from its confluence with the Columbia. Since his assign-ment had been to explore the northern part of Washing-ton, he decided to search other possibilities first. He established headquarters at Waterville. With C. B. Haskell, a fellow Great Northern engineer, as his assistant, he trav-eled the terrain between Grand Coulee and the Colum-bia. Then they examined the Columbia River from where it meets the Spokane River south to Pasco. Along the way they carefully examined every one of its tributaries, includ-ing the Sanpoil, Okanogan, and Methow. They went to Lake Chelan and traveled sixty miles from foot to head by

---

[20]John A. Gellatly, "History of Wenatchee," unpublished manuscript, 1958, 88.

rowboat. They continued on over the Cascade Pass, and about fifteen miles down the Skagit River on the west slope. Nothing seemed to favor routing the railroad that way to Bellingham.[21]

Next, Stevens and Haskell explored the Entiat River north to its head. Then they mapped out an imaginary route that went west from the Columbia to Cle Elum, across Snoqualmie Pass, and down Cedar River on the west side. Their diagram did not seem feasible. By then they had became thoroughly convinced that the Wenatchee River was the correct route to follow and turned their attention toward developing that possibility. They followed the Wenatchee River from its confluence with the Columbia through Tumwater canyon, and on up to Lake Wenatchee. They checked the summits of several small valleys, but none was satisfactory. From nearby Indian Pass they followed the crest of the Cascade Mountains all the way to Snoqualmie Pass.[22]

While going up the Wenatchee, Stevens observed what was afterwards known as Nason Creek and made a mental note of its location. He hiked along the top of the main mountain range and found a comparatively low place that had in it a creek flowing to the east. He felt certain it must be the beginning of the same creek he had noted earlier. He marked that gap with a blaze on a tree and continued his way along the range for a few days. After his return from that survey, he sent his assistant, Haskell, up the Wenatchee River to the mouth of that creek and then to its source. Stevens was sure the headwaters would be at the gap he had marked, and Haskell proved the theory correct. Con-

[21]Eva Anderson, *Rails Across the Cascades* (Wenatchee, Wash.: World Publishing Company, 1989), 15; and John F. Stevens, "Great Northern Railway," *Washington Historical Quarterly* 20 (1920): 111–12.

[22]Ibid.

On the Tum Tum, four miles west of
Leavenworth, Washington, work on the Great Northern
Railway. *Inland Empire Historical Society, Spokane, Washington.*

tinuing west, Haskell followed the early waters of the Skykomish River far enough to be sure he was over the Cascade Divide, then returned to the summit and carved "STEVENS' PASS" on a cedar tree. It was the second major mountain pass that Stevens had discovered within a year.

While Stevens was exploring the Cascades, a prospector wrote to Jim Hill telling him Stevens was wrong in looking for a pass in that area. He wrote there was a much better one that was overlooked and enclosed a sketch and description in the letter. Hill sent the map and letter to Stevens with orders to check it out at once. Frustrated because he realized no such pass existed, Stevens nevertheless carried out Hill's request. He found the prospector's camp and hiked up the creek shown on the drawing. After a day's search he concluded the pass was not in the main mountain at all, but through a side spur of it. Three days of his time had been wasted. It was a loss which he could ill afford.

Immediately after the pass was discovered and named for him, Stevens made a close study of it and its approaches, and decided that his search was ended. It lay in a favorable place and was relatively low in elevation at 4,000 feet. It afforded easy slopes for the construction of a switchback line, and a favorable site for a reasonably short tunnel if needed.

The work season was getting late, and it was imperative that a route be adopted by early spring. Stevens organized a party of engineers and put them to work making a hasty preliminary survey from the summit west, down the Skykomish River, where the escarpment of the Cascades is much more abrupt than on the east side. The work continued until deep snow on the mountain slopes made it impracticable to proceed farther. When Stevens got the results of the survey, he was bitterly disappointed.[23]

---

[23]"An Engineer's Recollections," 30; and Stevens, "Great Northern Railway," 112.

During the three winter months, he thought about and studied the line which the surveyors had put down on that part of the Cascades. He knew it was impracticable from its length, tortuous by the terrain, and prohibitively costly to construct. Yet, he felt that a suitable line was there somewhere to be developed. One night in March 1891, inspiration came to him, almost like a dream. Early the next morning he picked up his packer, John Maloney, and started for the mountains. They hiked all day and stopped that night at an abandoned lumberman's shack, the last place where shelter could be found. At daylight they continued on and reached a small valley similar to the one that Stevens had visualized two nights before. The line he created, which became known as the Martin Creek Loop, made it possible for rails to be laid through very rugged country.[24]

When snow melted enough in the spring of 1891 to make work possible, Stevens sent parties of engineers to survey the loop that he had sketched. Their results showed that it was a practicable route. As a temporary measure for getting over the pass, engineers charted a giant switchback, which is a zigzag way of getting down from a height too perpendicular to travel down headfirst. A tunnel was the preferred solution but the cost postponed its construction. The 2.6-mile-long Cascade Tunnel was started six years later, in 1897. It took three years to complete.

After Stevens discovered the pass and located a rail line over it, he sent a profile of it to Chief Engineer Beckler. In a brief time he received a letter saying that James Hill did not like one particular part of his location and was coming out to see about it. Stevens was directed that no work was to be done at that particular place. However, Stevens

---

[24]Ibid.

had a different idea that did not go along with that. When Hill, Beckler, and the contractors arrived to check on Stevens' work, he met them at the summit. It was Stevens' first contact with Jim Hill, who walked up to his location engineer with a single greeting: "Where is that thirteen-degree curve? I don't like it." Stevens replied, "Well, that's unfortunate, as the grading is finished, the bridge in, and everything ready for the track." Hill looked confused and said, "I want to see it." Stevens took the group to a point where they could get a view of the situation. Looking it over, Hill exclaimed, "That is all right, you could have done nothing else."[25]

James Hill approved Stevens Pass as the official route for the Great Northern Railway over the Cascade Mountains. Now the line for the Pacific Extension could be built between Spokane, at that time the western terminus of the tracks, and Wenatchee. During the summer of 1892, tracks were laid through the present towns of Wilson Creek, Ephrata, and Quincy to the Columbia River. They followed the east side of the river to a place called Rock Island. The cars were then ferried across the river on the steamship *Thomas L. Nixon* to tracks laid on the west side. (The next spring, 1893, a bridge was built at Rock Island for trains to cross the river.) Rails were laid along the west bank of the Columbia to Wenatchee, and continued on to Leavenworth and the pass.

While the Great Northern Extension pushed west from Spokane to Stevens Pass, the extremely difficult construction of the roadbed began in the Cascade Mountains. Hill, burdened with economic problems, had approved building the switchbacks as a temporary alternative to the tun-

---

[25]"An Engineer's Recollections," 32.

nel. Three switchbacks had been proposed for the east side of the range and five for the west, with grades set as steep as 4 percent. Because the entire area was covered with dense forest, crews had to clear the land so that engineers, workmen, and trainmen could see each other. The timber was used to build several trestles along the switchbacks.

With a work force of about 9,000 men and more than 3,500 teams of horses and mules, the construction of the Pacific Extension of the Great Northern Railway was completed in less than three years, 1891–93. The line extended from Havre, Montana, to Everett, Washington, a distance of 826 miles. The final spike was driven near the Cascade summit on June 18, 1893.[26]

That same summer, Stevens' assistant, surveyor C. B. Haskell, who named Stevens Pass, drowned when a small boat in which he was riding was swamped on the Columbia River.[27]

Changes came with completion of the Pacific Extension. Beckler resigned as chief engineer and left the railway. Hill appointed Stevens assistant chief engineer on the regular Great Northern staff with his office in Spokane. But the problems of developing traffic in a new territory, coupled with the depression of the mid-nineties, resulted in abnormally low profits for the Great Northern, and Stevens was among GN personnel laid off. For the last six months of 1894, he enjoyed a rest and vacation with his family in Everett. Then in December, answering Hill's summons to St. Paul, Stevens was surprised to be offered the job of chief engineer of the entire GN system.

Before he accepted, Stevens wanted to talk to N. D. Miller,

---

[26]Colley, "Stevens Has Blasted and Bridged His Way Across America," 88; and Anderson, *Rails Across the Cascades*, 23.

[27]Anderson, *Rails Across the Cascades*, 16.

who had succeeded Beckler as chief engineer. Having served under Miller, Stevens respected him professionally and personally, and was reluctant to replace him. Miller told him that change was inevitable, and he advised Stevens to take the job. On January 1, 1895, Stevens realized an ambition he had cherished for years—to become chief engineer of a great railroad. He was forty-one years old.[28]

His immediate action was to organize a capable department to push an extensive program of construction, improvements, and strong maintenance. He was able to hire competent men, delegate responsibility to them, and also keep tight control. His plans included upgrading stations, freight houses, and bridges, so he appointed an engineer of buildings and another of bridges. Resident engineers were located in St. Paul, Spokane, and Great Falls. Their number later increased.[29]

The first important construction projects authorized by the new chief were two railroad lines in northern Minnesota. Hill acquired railroad and iron ore-bearing land in the Mesabi Range for Great Northern stockholders in 1897. He also purchased an iron ore dock and a short railroad, the Duluth, Superior & Western. It extended from Deer River, Minnesota, to Lake Superior. The next year he acquired a short logging railroad, the Duluth, Mississippi River & Northern. It ran from Swan River to Hibbing, Minnesota. Stevens' assignment was to bring those two rough rail lines up to Great Northern standards and extend them. Building some two hundred miles across northern Minnesota was difficult, as drainage, grading, and rail laying were not adequate on the existing lines. Handling black

[28]Hidy and Hidy, *John Frank Stevens, Great Northern Engineer*, 352; and "An Engineer's Recollections," 32–33.

[29]Hidy and Hidy, *John Frank Stevens, Great Northern Engineer*, 352–53.

muck and bringing in suitable foundation material added to the construction problems.[30]

While working on the Duluth, Mississippi River & Northern line, Stevens found that the grade from Lake Superior to the summit of a hill surrounding it was 2 percent. Other railroads in that area all had a gradient of over 1 percent. Jim Hill wanted his line to have less than 1 percent and asked Stevens if he could get an eight-tenths line. The chief engineer replied that he thought there might be a six-tenths there. Hill sarcastically said, "You are trying to get sunshine out of cucumbers. An eight-tenths line is very good for that place." Stevens said no more and proceeded to try out a plan he had in mind. When the final location was finished, and maps and profiles prepared, he took them to Hill and told him he was ready to go ahead and build if Hill wanted it done. Studying the profile intently, Hill exclaimed, "Why, this is only a four-tenths grade!" To which Stevens slyly remarked, "Oh yes, just a bit of cucumber sunshine." With appreciation, Hill ordered the construction to begin right away.[31]

Another important construction under chief engineer Stevens was the first tunnel through the crest of the Cascades. Plans for the tunnel had been drawn in 1893, but it could not be built at that time. In addition to the proposed tunnel, a number of other improvements and changes were needed to make the main line over Stevens Pass more functional. The switchbacks over the summit had been installed as a temporary measure. Traffic over those thirteen miles was very expensive to operate, because trains had to be separated into two sections, and two engines put on each section. On each trip those engines burned a total of some three

---

[30]Ibid., 353–55.   [31]"An Engineer's Recollections," 34.

thousand pounds of coal. In addition, snow came early and stayed late on the crest of the mountains. Often snow fell at the rate of twelve inches per hour and formed twenty-feet-deep snowdrifts. Hundreds of men on both sides of the summit were hired to shovel out trains and snow plows.[32]

In 1896, Stevens reviewed the surveys made four years earlier for a tunnel. He made new plans, which Jim Hill approved and ordered the work to begin the following year. Hill didn't want to contract the project and gave Stevens the responsibility of being the builder. In 1897, construction of the tunnel began. The resident engineer, H. W. Edwards, was in charge, with M. E. Reed as his assistant. A. McIntosh was the general foreman, and his laborers were hard-rock miners and construction workers. The first part of the year was spent building living quarters, kitchens, dining halls, a hospital, and a powerhouse for the employees. The tunneling process began on August 20, 1897, from both ends.[33]

In determining the tunnel site, a rope—which would be above the tunnel with a valley between—was extended over two mountain peaks. The higher peak was on the west at 5,358-feet elevation. The lower eastern peak was at 2,150 feet. Length of the tunnel was figured by direct measurement rather than triangulation. A 400-foot steel tape was used to measure. Tension of the tape was regulated by a spring balance, and the temperature was observed at each end. Measurements were usually taken on cloudy days when there was little wind. The length of the tunnel was estimated to be 13,258 feet or 2.49 miles.[34]

For three years, eight hundred men worked in three shifts of eight hours each with no breaks, Sundays, or holidays.

[32]"Railroad Construction in Stevens' Pass," *The Confluence, Publication of North Central Washington Museum* 9, no. 2 (Summer 1992): 384–86; and Anderson, *Rails Across the Cascades,* 24–27.   [33]Ibid.   [34]Ibid.

As to be expected, there was a constant turnover of laborers. The crumbling nature of some of the soil created dust and slides. Abundance of snow and rain produced a great deal of water with which to deal. Prevailing winds, blasting fumes, and dust aggravated ventilation problems. The best modern machinery was used. Large drills from the west side hollowed through the mountains to join the bore of the drills from the east side, and a tunnel two and a half miles long emerged. Lined with timber and a wall of thick cement, the tunnel was twenty-three feet high, sixteen feet wide, and allowed for a single track.

On September 22, 1900, a final explosion was set off to blast away the last obstruction that separated the working gangs. General foreman McIntosh detonated the blast that destroyed the mound of earth which stood between workers from the west and east sides. When the smoke had cleared, loud shouts echoed along the tunnel, and congratulations were extended. The tunnel's alignment where the two gangs of workmen met was in error by less than one-fourth of an inch. The error in its grade was only two and one-fourth inches. The Cascade Tunnel was completed on October 13, 1900. Compared with the switchbacks, it cut off nine tortuous miles and eliminated 2,332 feet of curvature.[35] According to a Seattle newspaper, "The record of progress of work and the few casualties attending it has never been equaled, and the engineering work has never been approximated in the history of the world for similar construction."[36]

Other major projects accomplished by Stevens while chief engineer of Great Northern Railway included: relocating a rail line in northwestern Montana; constructing a 610-

---

[35] Ibid.

[36] *Seattle Post-Intelligencer*, Sunday, September 23, 1900, p. 10, quoted by Anderson, *Rails Across the Cascades*, 26.

203. Loop on the Great Northern R. R., near "Scenic," Washington.

Loop on the Great Northern Railway near Scenic, Washington.
Inland Empire Historical Society, Spokane, Washington.

mile line from Larimore, North Dakota, to Havre, Montana; acquiring the Spokane Falls & Northern Railway for the Great Northern and integrating it into the latter's system; improving all of Hill's railroads by raising tracks, reducing curves and grades, replacing 60-pound-per-yard rails with 75-pound and heavier ones; rebuilding wooden bridges and replacing some with steel structures; planning and starting new freight facilities; and constructing a tunnel to improve service and the approach to Seattle.

During the years of 1895 to 1903, much of the Great Northern Railway was rebuilt. Millions of dollars were spent for its development. Often Hill said to Stevens, "John, you must go ahead and fix up such and such a line." Sometimes the "fixing up" involved the building of many miles on a new location about which Hill knew nothing, so he left it all to Stevens' judgment.[37] Once there was a change of line that cost a large amount. Hill went over the old line on a special train to examine and see why the cost was so great. Then looking at the new line under construction, he said, "Look here, you are spending too much money over there." He was greatly disturbed. Stevens said nothing until their train reached a place where the whole project could be seen. He had the train stopped and asked Hill to take a look. He studied the situation for a few minutes and then exclaimed, "John, if you hadn't done this, I would have fired you. It would have been a crime to have spent money on this old line!"[38]

During 1898, Stevens became restless and felt the need for a change. He had been working day and night for three years. He went to Jim Hill and asked for a vacation. Hill told him he couldn't spare him at that time and refused to give him a vacation. Stevens immediately tendered his res-

---

[37]"An Engineer's Recollections," 33.   [38]Ibid., 34.

ignation, and Hill accepted it with ungracious comments which he concluded by saying, "You'll be back inside of a year." Hill then reappointed N. D. Miller, the former chief engineer, to take Stevens' place.[39]

Stevens went to British Columbia to forget everything for awhile. He contracted with the Canadian Pacific Railroad to build a heavy piece of mountain railway. However, his work was often interrupted when Hill would call to get his advice on engineering matters. After Stevens had been working ten months in Canada, Hill requested him to return to the Great Northern. For ethical reasons, Stevens talked to Miller about returning to the office of chief engineer. By then he felt rested and told Hill he would go back to St. Paul and resume his former job. He received a good increase in his former salary and an attractive bonus for returning.

James Hill had bought stock in the Chicago, Burlington, & Quincy Railroad and sent Stevens to inspect its main lines and evaluate the abilities of staff members who operated the railroad. Hill then went to Chicago to discuss the results of Stevens' study. While they were there, Hill offered Stevens the position of general manager of the Burlington Railroad. Stevens declined the offer because he didn't feel qualified for such an important position and realized there would be jealousy among the officials over a stranger from the Great Northern taking over the leadership. He told Hill the reasons for declining, but Hill waved them aside and insisted Stevens take the job. Stevens refused to change his decision, and after some heated words between them, Hill told his chief engineer he didn't care what he did or where he went and abruptly left. Stevens assumed his service for Hill had come to an end and returned to his office in St.

[39]Ibid., 35.

Paul. That quarrel was the only serious rift between those two men, and fortunately it was only temporary.[40]

A brief time after Hill returned to St. Paul, Stevens encountered him on the street. To his complete surprise, Hill asked him to become general manager of the Great Northern Railway. He then went on to remark, "I suppose that you will decline." Stevens was embarrassed because he didn't want the job, but he felt he must accept it. He told his boss he would take the position. On May 1, 1902, Stevens became Great Northern Railway's general manager, retaining also the office of chief engineer. The duties of the additional responsibility increased his work, but it all went smoothly.

Early in 1903, restlessness again settled within Stevens. His work no longer appealed to him as before, and he figured he had gone as far as he could with James J. Hill. He resigned both offices and went to work within another railroad. Years afterwards he stated that his reason for resigning was he felt Hill would have appointed him to the highest executive office of that vast company, and he didn't want it. "I felt there were and would be certain influences which would handicap me, and to counteract them would require the practice of a diplomacy which I was temperamentally unfit to exercise."[41]

Stevens had been with the Great Northern Railway at that time for seventeen years, including seven as chief engineer and two as general manager. The company had grown from a 3,000- to a 6,000-mile-long railroad system. It was said that Stevens had built as many miles of railroad as any other man in the world, and had built them economically and soundly.[42] James Hill, one of railroad's shrewdest builders,

---

[40]Ibid., 36.   [41]Ibid.

[42]John Fritz Board of Award, "Presentation of the John Fritz Gold Medal to John Frank Stevens," *Civil Engineering* (March 23, 1925): 25; and Colley, "Stevens Has Blasted and Bridged His Way Across America," 90.

aid of Stevens, "He is the most capable engineer in railroad construction that I have ever known. He is always in the right place at the right time, and does the right thing without asking questions about it."[43]

Stevens' reputation as a railroad builder became well-known. When he quit the Great Northern, he received numerous offers from other rail carriers. In 1903, the Chicago, Rock Island & Pacific line persuaded him to become their chief engineer. The following year they promoted him to second vice president. In 1905, the U.S. government asked him to leave the Rock Island road and lead a commission to the Philippine Islands to build railroads. However, the latter assignment never occurred as a new and greater one took Stevens to Panama.

---

[43]John Foster Carr, "The Panama Canal," Fifth Paper, "The Chief Engineer and His Work," *The Outlook* (June 2, 1906): 265,

4

CHIEF ENGINEER
OF THE PANAMA CANAL

In 1905, American construction of the Panama Canal, begun the previous year, almost came to a halt. The yellow fever epidemic, which first appeared in 1881, was taking a heavy toll among the workers both physically and psychologically. John F. Wallace, first chief engineer on the project, submitted his resignation on June 26, and the security of the entire program was threatened. President Theodore Roosevelt was deeply concerned and assigned William Howard Taft, secretary of war, and Theodore P. Shonts, chairman of the Isthmian Canal Commission, to begin searching for a replacement for Wallace. They examined the records of about twenty men and consulted leaders in transportation. James J. Hill told Taft that no man could be found in the entire country "better adapted" to build the Panama Canal than John F. Stevens.[1]

At that time, Stevens was in Chicago on his way to the Philippine Islands to supervise railroad construction for the government. President Roosevelt contacted him and asked him to serve as chief engineer of the Panama Canal project, replacing John Wallace. Stevens was reluctant and asked for time to consider the matter. A short time later, William

---

[1]Miles P. Du Val Jr., *And the Mountains Will Move, The Story of the Building of the Panama Canal* (Palo Alto: Stanford University Press, 1947), 183.

N. Cromwell, a lawyer who had been the chief negotiator in transfer of the Panama Canal project from French interests to the United States, called on Stevens. He outlined the embarrassing situation facing the Roosevelt administration, which feared public reaction in case of failure to build the canal. It was Cromwell's "persuasive tongue" that overcame Stevens' hesitation to take the position.[2]

Stevens also discussed the appointment with his wife, Harriet. Her prompt reply was, "Ever since you left Maine in 1874, you have been training yourself for this, the greatest engineering project in the world, and now it is offered to you. Please telephone at once and tell the President you will accept."[3] Stevens did, and Theodore Roosevelt appointed him chief engineer of the canal commission, effective July 1, 1905, with a salary of $30,000 for the year.

A tall, broad-shouldered man with steady, gray eyes, Stevens was fifty-two years old when he began his new venture. He had an open, swarthy, mustached face and looked squarely at the person to whom he spoke. He talked frankly and was deliberate, forceful, and intense, yet he spoke little unless he was reminiscing. He believed in a full day's work and had little time for trivial things.[4] This was the new chief engineer who was selected to carry on the construction of the Panama Canal.

*   *   *

The evolution of the colonization of the Isthmus of Panama is intriguing. It was discovered in 1501 by Rodrigo de Basti-

---

[2]Ibid.

[3]Tom H. Inkster, "John Frank Stevens, American Engineer," *Pacific Northwest Quarterly* 56, no. 2 (April 1965): 83.

[4]Carr, "The Panama Canal," 265.

das, who sailed with Christopher Columbus on his second voyage. In 1502, Columbus himself skirted the coast on his fourth voyage. King Ferdinand V of Spain sent Alonzo de Ojeda to be colonizer and governor of Colombia, and Diego de Nicuesa to do the same at Panama in 1509. None of those explorers was successful in establishing a colony due to native hostility, lack of provision, and disease.

The following year, 1510, Vasco Nunez de Balboa, an adventurer, stowed away aboard a ship that took him to de Ojeda's supposedly new colony at San Sebastian on the Colombian coast. He found de Ojeda had disappeared, and forty-one survivors were left. After a few months he induced them to transfer their colony to Panama, where it became the first permanent European settlement on the continent at Darien. It was named "Santa Maria La Antigua del Darien," commonly called Antigua. He persuaded the colonists to plant crops, taught them to befriend the Indians, and introduced many customs of the homeland. King Ferdinand appointed Balboa to be interim governor of the colony. The natives elected him to be their permanent governor.

Balboa believed that a long-sought "Other Sea" (the Pacific Ocean) might be in the area of the isthmus. When the natives told him there was a sea beyond the mountains to the south, he set out to discover it with an expedition of 190 Spaniards and 800 Indians. They crossed the isthmus and climbed a hill near the Gulf of San Miguel. There, in September 1513, Balboa saw the Pacific Ocean for the first time. In a few days they reached the Pacific, and Balboa claimed it and all its coasts for the kings of Castile. Impressed by Balboa's discovery, King Ferdinand named him governor of the "South Sea," Panama, and Coiba, an island off the Pacific coast of Panama.

The next year, 1514, the king appointed Don Pedro Arias de Avila, commonly known as Pedrarias Davila, to be governor of the Republic of Panama, with Balboa as his subordinate. He was envious of Balboa, and after several quarrels between them, he had Balboa killed by accusing Balboa of treason and having a friend give trumped-up evidence of it. Balboa was beheaded in January 1519. Governor Pedrarias Davila founded the city of Panama on August 15, 1519, and it became the seat of the government.[5]

Sir Francis Drake, an English adventurer, captured the settlements of Nombre Dios and Portobelo in 1572. The next year he intercepted the mule train carrying the Spanish king's share of treasure to that port from Panama and went back to England a rich man. Later, in 1595, he returned to Panama and burned the towns of Nombre de Dios and Santa Maria.

In 1671, an English pirate, Sir Henry Morgan, looted and burned Panama City. The ruins were abandoned, and two years later the natives rebuilt the present city five miles to the southwest.[6]

After three hundred years of Spanish control, Panama revolted. On November 21, 1821, the citizens adopted a resolution declaring the Isthmus of Panama to be independent of Spain, and then annexed themselves to the Republic of Colombia in South America. In 1845, Tomas Cipriano de Mosquera was elected president of Colombia and negotiated the building of the Panama Railroad across the isthmus with American financiers. It was fifty miles long, a

---

[5]Ira E. Bennett, *History of the Panama Canal* (Washington, D.C.: Historical Publishing Company, 1915), 23–47; *Encyclopedia Britannica* (Chicago: William Benton, 1969), vol. 2, pp. 383, 1067, and vol. 17, pp. 200–201; *The Encyclopedia Americana*, 1960 edition (Danbury, Conn.: Grollier, 1960), vol. 21, pp. 232–33.

[6]John F. Stevens, "The Panama Canal," *Transactions*, Paper No. 1650, American Society of Civil Engineers, New York, p. 948; *Encyclopedia Britannica*, vol. 2, pp. 383, 1067, and vol. 17, pp. 200–201; *The Encyclopedia Americana*, vol. 21, pp. 232–33.

single track of five-foot gauge, and had no sidings or station buildings. The gold rush to California in 1849 had
flooded the isthmus with men hurrying to mining locations
along the Pacific Coast. That pushed the construction of
the railroad. It was completed in 1855.

The hope of either finding a hidden strait connecting the
two oceans or building a canal to join them had been suggested as early as 1520, and was mentioned by Samuel de
Champlain also in the sixteenth century. Discussion about
it increased during the nineteenth century.[7]

Following a presentation of the subject at the worldwide
Congress of Geographical Sciences held in Paris in 1875, a
group formed in France with intentions to construct a sea-
level canal from Colón on the Atlantic Coast to Panama
on the Pacific. The group was known as La Compagnie
Universelle du Canal Interoceanique de Panama. The
leader and dominating force was Ferdinand de Lesseps, who
had previously constructed the Suez Canal in Egypt
(1859–69). De Lesseps' technical committee successfully
negotiated a contract with Colombia for land in Panama,
and on January 1, 1880, the French builders began the construction of a sea-level canal.

Before long they encountered problems that became
insurmountable: yellow fever (thought to be caused by poor
sanitation), a change of plans to build a lock canal instead
of a sea-level canal, expanded cost, and mismanagement.
The de Lesseps company was reorganized in October
1879, and continued operations on a limited scale until it
went bankrupt and was dissolved in 1889. The United
States was anxious to build the canal. Two viewpoints were

---

[7]Julian Hawthorne, *History of U.S. from 1492–1910*, vol. 3 (New York: P.F. Collier &
Son, 1910), 1137.

held as to where. Some believed it should be built across the isthmus at Nicaragua, and others felt the route should be at Panama. By the authority of Congress, President McKinley, in 1899, appointed a group of men known as the Isthmian Canal Commission. Their orders were to "investigate any and all practicable routes for a canal across the Isthmus of Panama, and also to investigate a proposed route across the southern end of Nicaragua, with a view of determining the most practicable and feasible place for such canal."[8] The commission pursued the work of investigation until the matter was closed by the United States' purchase of de Lesseps' rights and property at Panama. It was then the commission's responsibility to shape up the affairs, so that plans could be made and the actual work of construction started.

On the morning of May 4, 1904, the sale of the French property was finalized, with Lieutenant Mark Brooke of the U.S. Army taking possession for the United States. There was little ceremony, but the deeds were delivered. In February of that year President Roosevelt dismissed the original Isthmian Canal Commission. Its members had been unable to outline and carry on any progressive program of advancing the work, or offer any hope for the completion of any kind of canal at any location, in any reasonable length of time. The president appointed a new commission to replace the former one. Admiral John G. Walker was named the chairman. General George W. Davis, retired Army officer, was designated as governor of the Canal Zone. Other members were William B. Parsons, Benjamin M. Harrod, William H. Burr, and Carl E. Grunsky, all engineers, and Frank J. Heckler, a businessman. It

---

[8]Stevens, "The Panama Canal," 948.

was an Army, Navy, and civilian commission, with civilians in control of engineering.[9]

When the new commission met with the president at the White House on March 8, 1904, he told them they had been selected from the "best fitted" but warned that he expected a resignation from any member who should find the work "too exhausting and engrossing." He requested them to give special attention to sanitation and hygiene, using the best medical and sanitation experts. He wanted a rigorous supervision of expenditures and desired the best talent for every need. "What this nation will insist upon," he admonished, "is that the results be achieved."[10]

The commission set to the task of enlisting talented leadership. They selected Colonel William C. Gorgas, a distinguished Army doctor who had done outstanding work in Cuba as the chief sanitary officer for the Canal Zone. They arranged a trip to Panama. On the way they discussed who should be selected as the chief engineer, and decided to offer the position to John F. Wallace, one of the U.S.'s leading railroad engineers. The commission arrived at Colón on April 5, 1904. There they viewed the tremendous amount of preparation that needed to be done before actual work on the canal could begin. Although the United States originally contemplated a lock canal, a number of employees in Panama urged the commission to investigate further both the practicability of a sea-level canal as well as to determine the summit level and lock data for a lock canal. The committee stayed in Panama two weeks using the home built for the former French engineer de Lesseps. Then they returned to the United States to organize engineering groups.

---

[9]Du Val, *And the Mountains Will Move.*
[10]Ibid.

On May 6, John Wallace was formally offered the position of chief engineer at an annual salary of $25,000. He accepted the position effective June 1, 1904, and agreed to maintain his residence on the isthmus. The survey-engineer groups went to the isthmus to begin their work. They had to go into the jungle, and their supplies were so scarce that they were compelled to live like the natives. The only way they could get fresh meat was to kill monkeys. Some of them stated that the carcass of a skinned monkey so closely resembled that of a human baby that they could hardly eat it.[11]

General Davis and Colonel Gorgas went to Panama with the surveying parties. General Davis set about organizing a civil government of the Canal Zone. He issued a proclamation of his governship, and then negotiated an agreement with the Panamanian government concerning the relationship between them. He also created a sanitation department and appointed Colonel Gorgas to be in charge.

Chief Engineer Wallace arrived in Panama the last part of June 1904. His first project was to sort out and get into shape the equipment which the French had left. From one side of the isthmus to the other were rows of houses, machinery, material, and junk. There were 2,148 French buildings, including the Ancón Hospital, the administration building in Panama City, the Tobago Sanatorium, and the residence of the director general. Workers assigned to prepare the isthmus for the canal diggers had to occupy them until new ones could be built. The railroad had been left as mere lines of rust and decay. The rolling stock of engines and cars had degenerated into an almost useless state.

---

[11]Bennett, *History of the Panama Canal*, 127.

By equipping the machine shops, Wallace was able to repair five or six of the French locomotives and about one hundred dump cars a month. He put the men to work at once digging the Culebra Cut. Work on the cut—a canyon dug out of the backbone of the intercontinental divide on the Cordilleras Mountain Range—was begun by the French. It extended from Lake Gatun for nine miles southeast toward the Pacific. Its bottom width was to be 200 feet but was later changed to 300, with an average depth of 120 feet and a top width of one-third of a mile in places. It was to lift up the waters so they could meet part of the way above the oceans. Before long, Wallace was in trouble with members of the Isthmian Canal Commission, who resented his authority. He lost patience with the government auditing system in Washington because of its slow action. Requisitions for material and equipment took a long time arriving. There were no acceptable quarters nor suitable food supplies for the workers. Yellow fever was spreading. By the end of the year, there were internal dissensions within the commission.[12]

Wallace went to Washington to confer with President Roosevelt. As a result the president ordered the commission to resign, and he appointed another new one. It consisted of Theodore P. Shonts, chairman; Charles E. Magoon, the new governor of the Canal Zone; John F. Wallace, chief engineer; Mordecai T. Endicott, Peter C. Hains, Oswald H. Ernst, and Benjamin M. Harrod. He also reorganized the Panama Railroad and placed it under the virtual control of the chief engineer.

After returning to Panama, Wallace personally felt the

---

[12]Du Val, *And the Mountains Will Move*, 133; Bennett, *History of the Panama Canal*, 126–28.

distress of the raging yellow fever epidemic when it took the life of his secretary's wife. In a short time he called William Taft, secretary of war, for permission to return to the States and discuss a personal problem. Secretary Taft granted his request. When the two came together, Wallace told Taft about his disgust with red tape, of his fear that he or his wife might get yellow fever, and that he had been offered employment elsewhere with a much better salary, between $50,000 and $60,000. Taft became angry over that information and there were some heated words between the two men. Finally, Taft said, "If you are going to resign at all you might as well resign now." Wallace wrote his resignation to President Roosevelt, and the president accepted it on June 28, 1905, effective "immediately."[13]

John Wallace's resignation was an undeniable blow to the project. It came at a critical time in the history of the canal, and it came in a way that demoralized the work force on the isthmus and shook public confidence at home. Chaos and hysteria seemed to threaten the security of the canal program which was still in the experimental and development stage.[14]

*   *   *

Just two days after Wallace submitted his resignation as chief engineer of the Panama Canal, President Roosevelt offered the position to John Stevens. As mentioned earlier, he was reluctant to accept it. He finally succumbed to the persuasive tongue of the lawyer-negotiator, William Cromwell. He had presented a doleful picture of the dis-

[13]Du Val, *And the Mountains Will Move*, 169–74.
[14]Hon. Daniel J. Flood, House of Representatives, "Anniversary of decision to build lock-type canal and compliments to John F. Stevens," *Congressional Record*, House, 84th Congress, 2nd session, May 29, 1956, p. 9285.

couraging condition into which affairs had drifted during the American occupation of the isthmus. Then he challenged Stevens to accept the chief engineer's responsibilities as a loyal citizen and supporter of an administration much perplexed over the situation. Stevens consented with a verbal agreement. Shonts telegraphed Stevens his appointment as chief engineer at an annual salary of $30,000, effective July 1. Stevens accepted it on June 30.[15]

He then called upon President Roosevelt at his home in Oyster Bay, New York. The president told him that the affairs at Panama were in a "devil of a mess." Stevens took the opportunity to state the conditions on which he would accept the position of chief engineer. The first was that, while technically he would be under the orders of the commission, he was to have a free hand in all matters, and would be hampered by no one in authority, high or low. The second was that he would not agree to remain until the canal was finished, but would remain until the failure or success of the project was assured according to his judgment. President Roosevelt made no objections to those conditions. He said that the canal must be built, and that he depended on the new chief engineer to make it a success. He refrained from giving any orders as to how matters should be handled.[16]

The president, in his characteristic manner, told Stevens an anecdote: A certain man suddenly became wealthy and set up a large establishment as a home. When his butler arrived he said to him, "I don't know in the least what you are to do—but one thing I do know, you get busy and buttle like hell!" The president also advised Stevens that the latter should address him directly in matters concerning the work. When Stevens wondered if such action might raise

[15]Bennett, *History of the Panama Canal*, 210–224.   [16]Ibid., 949–50.

conflicts with Secretary of War Taft or the Isthmian Canal
Commission, Roosevelt brushed the objection aside by stating that all parties involved understood his wishes in the
matter.[17]

The appointment had happened so quickly that Stevens
was unable to think much about planning before the press
began to seek interviews. His general statements to them
were that he realized the vastness of the undertaking and
that it was the "greatest of its kind in history," but he hoped
to carry out his part.[18] He expected to establish his family
residence on the isthmus and dedicate all his efforts to construction of the canal.

He had read extensively about the conditions on the isthmus, and the Panama dangers were no problem to him. He
could say, "For three years in Mexico I stood the test of chills
and fever incident to malaria. I have slept under wet skies
on the Western plains, rolled only in a single blanket, and
I have experienced the rigors of a far Northern winter under
primitive conditions."[19] To the rugged Stevens, tropical
Panama could be no worse than the Philippines, where he
would have been had the Panama opportunity not presented
itself.

Early in July he went to Washington, D.C., and organized a corps of civil engineers to supervise various areas of
work. He hired on the basis of fitness, which was unpop-
ular in certain circles. The leader of the American Federation of Labor tried to compel him to make a closed shop.
Stevens refused, and the labor leader threatened to take the

---

[17]Du Val, *And the Mountains Will Move*, 185; Flood, *Congressional Record*, House, pp.
9285–9289.
[18]Du Val, *And the Mountains Will Move*, 184, quoting from *Panama Star and Herald*,
October 22, 1905.
[19]Ibid.

matter to the president. Stevens told him "he could take it to the Lord, if he chose, but it would make no difference, and to close the door when he went out."[20] He went out at once, and nothing further was heard on that matter. In planning for the food situation on the isthmus, Stevens arranged a contract with a Chicago firm for fresh meat to be sent via Panama Railroad steamers at a cost of about 60 percent of Chicago retail prices.

Satisfied with information he had received and looking forward to the vast task set before him, Stevens boarded the ship that would take him to Panama. He sailed along with commission chairman Shonts. Neither one had ever been to Panama. Secretary Taft had cabled to Panama governor Magoon: "Shonts and Stevens will soon be with you, and the mountain will move."[21] Flying dirt or moving mountain were insignificant to the tremendous necessity that the United States' venture in Panama be rescued from a humiliating defeat. So far it had failed miserably, and the *New York Times* reported that it had already spent some $128 million. Much talk had been about French inefficiency and failure versus American efficiency and know-how, but thus far the U.S. had performed with less efficiency, purpose, and courage than the French. A whole year had been lost, and the situation on the isthmus was in shambles.[22] The government didn't even know what type of canal it wanted to build.

Stevens and Shonts arrived at the Colón harbor on July 25, 1905. Because the isthmus had just passed the height of a yellow fever epidemic, there was no spectacular reception for the new officers, and they went directly to the home of

---

[20]Stevens, "The Panama Canal," 950.

[21]David McCullough, *The Path Between the Seas* (New York: Simon & Schuster, 1977), 462. [22]Ibid.

Governor Magoon. Later, Stevens wrote that conditions could have been much worse, but they were bad enough. No real start had been made at any effective work on the canal proper—no adequate organization had been observed, sanitary reforms were just beginning, and only a small amount of supplies that were absolutely necessary had been ordered. There was no cooperation apparent in the organization that did exist, and as far as he could discover, no systematic plans had been made to carry out the work with any degree of success. Probably worst of all was that among the disjointed force of white employees hovered the angel of death in the shape of yellow fever. Stevens later wrote,

> No one will ever know, no one can realize, the call on mind and body which was made upon a few for weary months, while all the necessary preliminary work was being planned and carried forward, and no attempt was or could be made to carry on actual construction until such preliminaries were well in hand.[23]

After dinner in the home of Governor Magoon on the first evening after his arrival, Stevens joined Shonts, Magoon, and the chief sanitary officer, Colonel Gorgas, to discuss the future of the Panama Canal. They concluded that the most urgent needs were the housing and feeding of employees and the eradication of yellow fever. There had to be homes, markets, entertainment, and good supply lines to keep the working forces contented and in good health. A few days later, Shonts told Stevens that Gorgas would have to be relieved and replaced by a younger medic who could speed up the sanitary program. Stevens strongly objected to that, and Gorgas was retained. The chief engineer knew the importance of sanitation and overcoming yellow fever. When he organized his staff of engineers, he

---

[23]Bennett, *History of the Panama Canal*, 211.

John Stevens, chief engineer, at his desk in Panama.

assigned almost all of a division to Gorgas for sanitation work. The rest of the engineers were set to work on repair and construction of housing.

For nearly a week Stevens and Shonts studied the terrain and the general situation at the Canal Zone. Then on August 1, Shonts returned to Washington and left Stevens in full charge of construction on the isthmus. He quickly proved himself to be a leader of action as well as of ideas. With an imposing stature and commanding personality, he energetically walked the entire length of the canal, checking the work projects. Often speaking to the employees, he told them there were only three diseases on the isthmus—yellow fever, malaria, and cold feet, and the great-

Stevens and an employee on walkway at Panama Canal.

est of these was cold feet.[24] Stevens displayed no air of
authority. He never used a special train to transport himself around the area. He had brought no cronies down for
fat jobs. The men he selected had a way of proving their
ability. Each worker he discharged deserved dismissal by
almost common consent. He vetoed the proposal to build
himself an expensive home at Ancón, and designed a

[24]John Foster Carr, "The Chief Engineer and His Work," *The Outlook* (June 2, 1906):
265–66; Flood, *Congressional Record*, House, pp. 9285–9289.

cheap home in which to live on the Culebra Heights, where the most important work lay. He was a hard taskmaster but worked hard himself. A claim around the Canal Zone was that one could "take a spy glass, and up or down the road you'll see Stevens striding over the ties."[25]

A carpenter had been told to build some sheds at Gorgona. He wrote back to Stevens to say that there was some old French equipment in the way, and asking what he should do. Stevens scribbled on the bottom of the letter, "Wait 'til I have a free Sunday and I'll come down and move it for you."[26] He had reproofs for dodgers of responsibilities, but he often showed kindness. Once a young engineer gave him an unasked-for report on a difficult job that had been given him to do. Stevens told him, "You are not hired to advise me how your work can be done. You are hired to do it. You may make mistakes, but you were appointed because I know you are equal to your work. And remember this—there is only one mistake you can make that will be absolutely fatal with me, and that is to do nothing."[27]

Stevens was aware that before his appointment as chief engineer, President Roosevelt had been anxious to decide what kind of canal should be built. He had appointed an international board of engineers consisting of four Europeans and four Americans, all experts in their profession. They were to examine conditions, collect data, call witnesses, and attempt to select the style of canal best suited for the site.[28]

A sea-level canal would be like a huge canal-ditch extending across the isthmus from the Atlantic to the Pacific Ocean. It would be 49 miles in length, with 39 miles of that

[25]Carr, "The Panama Canal," 266.   [26]Ibid.
[27]Ibid.   [28]"An Engineer's Recollections," 40.

distance less than 200 feet in width. Three-fourths of the
excavation would be through rock under the water. Traffic
through it would be very slow because when two ships met,
one would have to tie up at a station on the bank to let the
other pass. Further, it would require the entire capacity of
such a ditch to move back and forth the many tugs,
dredges, and cranes that would be necessary just to keep it
open. Also, there would be the problem of what to do with
the Chagres River, a wild, torrential river located in one of
the heavier rainfall areas of the world in the northwest Isthmus of Panama. The river flowed southwest and often
flooded, making it extremely hard to control. It was nearly
impossible to connect it safely with a huge canal-ditch.
Stevens estimated the cost would be $150 million more than
a high-level lock canal.

A high-level lock canal would require that a huge dam
be constructed at Gatun on the Atlantic side to catch the
Chagres River and form a lake to become part of the canal.
The plan for the Gatun site had been proposed by the
French engineer Godin de Lepinay when French interests
had considered building the canal. Gatun Dam would
maintain a head of 85 feet of water. Its height would be
reduced by a double flight of locks with 43 feet of lift each,
and the reduced level would continue to the vicinity of Pedro
Miguel on the Pacific side through the Culebra Cut. Making the cut would prove to be an amazing accomplishment—a canyon dug across the backbone of the Cordillera
Mountains, the intercontinental divide that separates the
East from the West in both Latin and South America. It
would be nine miles long, with an average depth of 120 feet,
a bottom width of 300 feet, and a top width that would
reach one-third of a mile in places. At Pedro Miguel, dou-

ble-lift locks would drop the water to sea level on the Pacific Ocean. A tide-regulating lock would be installed in the vicinity of Miraflores.[29]

After a week of surveying the situation, Stevens figured the best choice would be an 85-foot-high lock canal. Admittedly, at first, he had been inclined toward the popular idea of a sea-level canal, but after observing the topography in the light of operational and navigational needs as well as engineering and construction demands, he decided on the high-level lake plan. When the international board of engineers reported their conclusions to him, all four foreign members and two of the Americans had voted for the sea-level canal. The other two Americans made a minority report in favor of a lock-type. Stevens rejected the majority's decision.

However, with its attractive title, the sea-level style of canal seemed to be preferred by the public. For a large number of the public, news media, and politicians, the idea of a sea-level canal meant digging a huge ditch across isthmus. It seemed this would be much quicker and cheaper to build; therefore, that type of canal was attractive to them. After a majority of the international board had reported in favor of such a type, the matter appeared to be settled. Yet, some of the most competent engineers who had given the matter close study, such as Stevens and the two who had presented the minority report of the international board, were of the opinion that the lock-type was definitely the correct solution. It would be the means of controlling the Chagres River. That group became zealous promoters of that kind of canal.

As Stevens continued to work on setting his priorities and

---

[29]Bennett, *History of the Panama Canal*, 146, 149.

methods for constructing the canal, he felt handicapped because he didn't know which type of lock would be selected because the government hadn't made any decision. In early December 1905, he wrote to the Isthmian Canal Commission stating his conclusion. He elaborated on the types of the canal, emphasizing that under the conditions that existed, the construction of a sea-level canal was not practicable.

The Isthmian Canal Commission accepted the chief engineer's report on January 26, 1906. They added some arguments in favor of it and sent it on to President Roosevelt, strongly recommending the lock-type canal. The president forwarded the report to the Congress on February 19, 1906, strongly supporting the high-level lake plan. Stevens was called to Washington to testify, and deftly voiced his reasons for his choice of canal. The House quickly approved the recommendation, but the Senate delayed voting on it. Congress considered the matter again in June, and Stevens had to return to Washington to testify. This time he found President Roosevelt cool to the idea of a lock canal and leaning toward the sea-level canal. Stevens was accustomed to handling superiors as well as subordinates in a vigorous manner. He went to the president and talked to him like a Dutch uncle. Whatever the chief engineer said to the president, it convinced him again that Stevens' plan was the correct one.

In 1956, the annual meeting of the Panama Canal Society of Washington, D.C., was the fiftieth anniversary of the decision to select the lock-style canal for Panama. Events of that time were recalled by Stevens' son, John F. Stevens Jr. He wrote:

> In June 1906, father was in Washington at our home, and I was home from school. He learned from the press that a decision had

been reached on the recommendations of the International Board of Engineers appointed by President Theodore Roosevelt, and that it was decided to build a sea-level canal.

When father read this news, he telephoned the White House and asked for an appointment with the President for that very evening, which was given him for 8 o'clock.

When he returned home and came through the hall, he tossed his hat over on the table and said: "Well, that's settled. There will be no sea-level canal built at Panama."[30]

A few days after their meeting, President Roosevelt insisted in the Congress on a lock canal, which was then designed and built.

With the support of President Roosevelt, Secretary of War Taft, and the Isthmian Canal Commission, the chief engineer continued to lobby for the high-level lock canal. He was called to testify before the Senate committee. For two hot days in June he argued the merits of the lock-type canal. One senator asked him off the record if he seriously meant to tell the senators that an earth dam 100 feet high would hold back a lake thirty miles long. Stevens replied, "Well, Senator, the eastern dikes of Holland hold back the Atlantic Ocean."[31]

The debate on floor of the Senate was led by Senator Philander C. Knox, who championed the lock-type using arguments and refutations prepared for him by Stevens. When it came time to vote, the Senate passed the lock-type canal by a small majority; in his recollections, Stevens wrote that the difference was seven votes.[32] On June 29, 1906, Congress made the momentous decision regarding the Panama

---

[30]Flood, *Congressional Record*, House, p. 9286. At Congressman Flood's invitation, a letter was read by Earl Harding, Brooklyn, New York, from John F. Stevens, Jr., written before his recent death, citing information about his father.

[31]"An Engineer's Recollections," 42.

[32]Flood, *Congressional Record*, House, p. 9288.

Canal. It adopted the high-level lake and lock plan so strongly recommended by Stevens. The next day, Roosevelt showed his confidence in Stevens by appointing him as a member of the Isthmian Canal Commission in addition to his position as chief engineer.

Stevens had spent a good amount of time in Washington away from his work, and his enemies began putting pressure on him. Those who opposed building any canal sought means to delay its construction. Others who advocated the sea-level canal were stung in their defeat by Congress's action, and were waiting to strike at any change in the approved plans. Changes would indicate weaknesses in the high-level lock canal. Together, those two factions represented a political and economic power that the chief engineer couldn't ignore.

He returned to Panama on July 4, 1906, still concentrating on his building plans. He felt some corrections were needed for the Pacific locks. They needed to be spaced out more. By August 3, he had settled on a plan to place all Pacific locks in one group of three lifts south of Miraflores with the terminal dam and locks between two hills, Certo Aguadulce on the west side of the canal and Cerro de Puente on the east. That location would provide the same lock arrangements at both ends of the canal, avoid creating a traffic choke at Pedro Miguel, enable uninterrupted summit-lake navigation from the Atlantic locks to the Pacific, and supply a lake-level traffic mobilization anchorage at the Pacific end to match that at the Atlantic.[33]

He was anxious to begin construction. Beside building the Gatun Dam, he knew the other key project was excavating the Culebra Cut. Over 100 million cubic yards of

---

[33]Ibid.

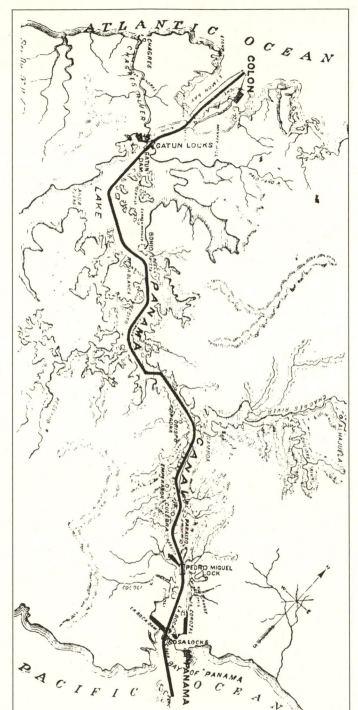

Panama Canal.

material had to be hauled away and dumped. From his experience, he realized the job could be done by steam shovels and work trains, but the main problem to overcome was transportation—rock and earth could be excavated quickly, but the transportation challenge of hauling them away would take longer to solve. If steam shovels were to do the work, their buckets had to be kept moving every possible minute.

Stevens devised a system of railway tracks that permitted the minimum loss of time in moving empty train cars to the steam shovels, and in taking away the loaded ones. He selected new dump sites and changed the old ones in such a way that trains could be unloaded in the shortest possible time. His system was so successful that it continued until the entire mass of material had been removed from the cut.[34]

Plans and basic projects had to begin and be completed in order to make the Canal Zone a fit place in which to live and work. Stevens' vast crew of engineers had mountains of work ahead of them. Construction areas for building the canal had to be determined and assigned. The isthmus had to be sanitized and yellow fever eliminated. Housing had to be provided for all classes of employees, from department heads to laborers. Food supplies that needed to be obtained would allow all employees an opportunity to get fresh meat and vegetables at reasonable cost.

Stevens moved the headquarters of the engineering organization from Panama City, where it was occupying the buildings of the old French de Lesseps regime, to a new town that was built at Culebra, a point about midway of the

---

[34]Theodore P. Shonts, "The Railroad Men at Panama," in Bennett, *History of the Panama Canal*, 202–9.

famous cut. They had a spacious building with offices for heads of the various departments. An extensive telephone system was installed, covering the entire Canal Zone, Panama, and Colón. Sitting at his office desk, the chief engineer could contact without undue delay every office, shop, steam shovel, and switch shanty along the entire line.

With his assistant, J. G. Sullivan, Stevens brought in W. G. Bierd, assistant general manager of the Chicago & Rock Island Railroad, to take charge as general superintendent of the Panama Railroad. Tons of freight were piled in cars and warehouses. Some of the shipments had been lying there for over a year. That congestion had to be cleared out. They then double-tracked the railroad all the way across the isthmus, except over Culebra Hill and from Mount Hope to Gatun, a total of 47.11 miles. They increased the dock capacity at both Cristobal and LaBoca, replaced the light 56-pound rail with 70-pound rail, and strengthened the bridges along the line to enable them to carry heavy loads. They built new stations and signals, a telegraph line, and purchased larger passenger and freight cars and locomotives. They equipped the railroad steamers with refrigerating plants, erected a cold storage plant at Colón, and ordered refrigerator cars for immediate shipment. Thus a line of refrigeration was established from the markets of the United States to commissaries being built along the canal, and the employees could enjoy fresh produce in their meals.[35]

During the summer of 1906, diseases on the isthmus continued to be a threatening problem. Yellow fever, malaria, pneumonia, tuberculosis, and intestinal illness were still

---

[35]John Fritz Board of Award, "Presentation of the John Fritz Gold Medal to John Frank Stevens," 29; Shonts, "Railroad Men At Panama," in Bennett, *History of the Panama Canal*, 207.

rampant. In Washington, D.C., Theodore Shonts and William Taft, head of the Isthmian Affairs, continued a campaign to have Colonel Gorgas removed as chief sanitary officer. Shonts had little confidence in Gorgas' theory that mosquitoes caused yellow fever. Also, he felt that the sanitation efforts of Panama should be a high priority. Taft believed that Gorgas had "no executive ability at all."[36]

At first, Stevens was doubtful about the mosquito theory, but since Gorgas was there on the job and responsible for making Panama a healthy place to work, he felt it was his duty to back Gorgas to the fullest. Shonts had located a replacement for Gorgas, a man named Hamilton Wright, and gave his name to President Roosevelt. Stevens learned of Shonts' action and fired a letter to the president in which he insisted that Gorgas remain as the sanitation officer. Once again the decision on a vital issue was left ultimately to the president.

Roosevelt consulted two friends about the situation: Dr. William H. Welch of Johns Hopkins Medical School, and Dr. Alexander Lambert, a friend and hunting companion. When answering Roosevelt's request for a recommendation on Gorgas, Welch replied that no one was better equipped for the work than Gorgas—the best man was already on the job. In a private conversation with the president, Lambert said,

> Smells and filth, Mr. President, have nothing to do with either the malaria or the yellow fever. You are facing one of the greatest decisions of your career. You must choose between Shonts and Gorgas. If you fall back upon the old methods of sanitation, you will fail, just as the French failed. If you back up Gorgas and his

---

[36]McCullough, *The Path Between the Seas*, 465.

ideas and let him pursue his campaign against the mosquitoes, you will get your canal.[37]

On the positive advice of those two friends, President Roosevelt decided that Gorgas should stay. He summoned Shonts to the White House and told him to support Gorgas. Shonts did, and saw to it that the sanitation department got what it needed.

Chief Engineer Stevens had placed a division of his engineers under the direction of Colonel Gorgas. Within a few months the sanitation officer had 3,500 workers. To exterminate the mosquitoes, a drastic program for destroying their breeding places was inaugurated. The workers cleared underbrush and vegetation from every area where there was any work or habitation for a distance of 1,000 feet, and they drained standing water or covered it with oil. Any receptacles that might retain water were removed or covered. They cleaned up rubbish piles and fumigated time and again the towns, villages, and labor camps in the Canal Zone. They also fumigated residences house by house, and worked by units, one city, village, or camp at a time.

In Panama City the water supply was from shallow wells located throughout the town, into which sewage sometimes seeped. The engineers built a concrete dam across a mountain stream twelve miles away. From that reservoir clean water flowed through a large iron pipe to a cement distributing pond on the Ancón hills. From there, gravity carried water to all parts of Panama City. They installed a water line to each house. The sewage problem was solved by filling the old wells with dirt and building a new system that discharged its contents into the bay. Heavy tides washed away the waste. They removed the old surface from

---

[37]Ibid., 467–68.

Panama's main streets and re-paved them with vitrified brick laid on a cement foundation. They macadamized the less important streets.

For the new towns of Colón and Cristobal, workers built a large earthen dam back in the hills and laid a large cast iron pipe to carry water to them from the reservoir. In both communities they put in water mains that covered the entire areas. Every house, office, and shop had its own water supply. Installing a sewerage system for the two towns was a difficult project. Since they both were situated on coral reefs, the natural elevation was less than three feet above sea level. The workers solved the problem by constructing a huge underground concrete-lined reservoir. All sewer lines discharged their contents into it. Electric pumps then pumped the sewage through a drain pipe that extended far out into the bay. It was a very adequate and workable system. Engineers also graded the streets of Colón and Cristobal and paved them with brick and built sidewalks along the streets.

The sanitation crusade included the building of hospitals at Colón and Ancón, and a number of smaller ones along the line of the canal. Observance of sanitary laws was strictly enforced. If an employee was discovered with a high fever, he was compelled to go to the hospital, whether he wanted to or not. In less than four months yellow fever was exterminated.

Much praise was given to Gorgas, his staff, and engineer laborers for their notable work in getting rid of diseases and making the Canal Zone a safe place in which to work. Yet, the real hero, in Gorgas' opinion, was John F. Stevens. In a private letter to Stevens years later he wrote,

> The fact is that you are the only one of the higher officials on the isthmus who always supported the sanitary department . . .

both before and after your time. So you can understand that our relations, yours and mine, stand out in my memory . . . as a green and pleasant oasis.[38]

Simultaneously with the engineers working on sanitation, another division labored under W. M. Belding on the construction of buildings. They renovated hundreds of houses taken over from the French and built new dwelling houses and living quarters. They designed and constructed hotels, restaurants, clubhouses, schoolhouses, courthouses, post offices, jails, commissary buildings, and fire engine houses. Along the line of the canal they built a number of villages containing populations ranging from a few hundred up to five thousand each. In each one of the central towns they built a clubhouse or recreation building for the employees. Each one contained a gymnasium, reception room, card room, billiard room, and an assembly hall.

Stevens assigned another group of engineers to be the department of labor and quarters under the leadership of Jackson Smith. Their responsibilities were to recruit skilled and unskilled laborers and to house and feed them. Skilled labor was obtained from the United States. Agents for that purpose were placed in several large cities that were centers of manufacturing and railway activities. They carefully selected 3,243 workers, but many of the men discovered they didn't like living in Panama, and, within a year, more than half had quit and returned home.

Recruiting unskilled laborers adequate to build the canal was a very difficult task. Stevens at first thought about getting Chinese workers since he had observed their good qualities in building railways and similar works on the Pacific Coast. However, after careful consideration, the isthmus

---

[38]Stevens, "The Panama Canal," 953.

commission had good reasons for dismissing that idea:the French had failed to build a canal, and the Chinese government refused to permit their workers to go to Panama. It was suggested that Spain offered a good supply, so the chief executive sent a recruiting agent there. Nearly eight thousand workers were brought to the isthmus from Spain's Biscayan provinces. Agents also recruited workers from Italy, Greece, and the West Indies. Laborers from the West Indian islands proved to be slow and indifferent. When Stevens established the wage scale that continued to the completion of the canal, he granted twenty cents an hour for unskilled white laborers and ten cents an hour for black laborers.[39] After a few days of work, the Spaniards went on strike and tried to force the black workers out of the Culebra Cut. A force of the Canal Police quickly appeared, and a conflict took place. Several of the Spaniards met immortality, and the majority went back to work. No more trouble was had from them. Stevens often went out along the line, checked the various projects, traveled in overalls through jungle and over hills, rode the trains, and talked with the workers. He walked with energy and was confident. He was the type of person other men naturally looked to as a leader and whose presence is felt.[40]

Finally the construction of buildings was almost completed. About 75 million feet of lumber, mostly from the Puget Sound area in Washington State, was used in eighteen months. To finish such a vast amount of construction in twelve years was amazing. The interesting feature is that it was accomplished by Stevens' civil engineers and not by the Army Corps of Engineers, as some later critics mis-

---

[39]Bennett, *History of the Panama Canal*, 206–7, 213–14.
[40]DuVal, *And the Mountain Will Move*, 191–92, citing Farnham Bishop, *Panama, Past and Present*, 159.

takenly believed. It was built in a wild and unproductive country two thousand miles away from home, and most items had to be requisitioned and shipped from American factories.[41]

Every employee of every grade, black or white, was furnished a house with free rent, lights, and fuel. Married men were encouraged to send for their wives and families. To avoid quarrels or rivalry over accommodations, a rule was made that each man should get one square foot for every dollar of his monthly pay. That applied to both bachelor and married quarters. Wives were also entitled to a square foot per dollar earned by their husbands. A different rule applied to children. The housing rule was devised and enforced by Jackson Smith, who was known afterward as "Square Foot Smith."[42]

To feed the huge crowd of workers and their families took skillful planning. For some years the Panama Railroad had maintained a small commissary through which its employees could obtain necessary food. Hardly any food was grown locally because of crop failure the two previous years, and there was no foodstuff to be shipped in from neighboring provinces. Everything had to come from the United States. Stevens and the isthmus commission decided to continue the commissary arrangement of the railroad at Cristobal. They had it enlarged and developed into an immense department store to carry nearly every article of human need. Supplies were sold at low prices by a system of cash payment through coupons.

In Panama at that time was an experienced railroad mess hall operator named Jacob E. Markel. He was looking for

[41]Bennett, *History of the Panama Canal,* 211, and Stevens, "The Panama Canal," 959.
[42]McCullough, *The Path Between the Seas,* 477–78, citing R.E. Wood, "The Working Force of the Panama Canal," in Goethals, ed., *Panama Canal,* vol. 1, 199.

a contract to feed all canal forces. He left the isthmus and went to Washington, D.C., to lobby the Isthmian Canal Commission for a contract to feed employees of the canal. The commission gave the idea serious attention and called for bids. Markel was low bidder at $36 a month per laborer. When Stevens received the news, he and his employees exploded in great dissatisfaction. The bid would necessitate an increase of pay, plus the contractor would be getting a million dollars a year clear profit. The chief engineer protested the contract and it was canceled with the consent of Markel, who stated it would be impossible to do the work with Stevens "cutting his bowels out."[43]

Since the commission in Washington hadn't come up with an adequate plan for feeding the employees, Stevens continued to take matters in his own hands. In every large labor camp he opened commissary stores that carried fresh meat and groceries for married workers and others who wanted to prepare their own meals. Eating places were provided for single workers. Both the meals and the groceries were sold at cost. Under Jackson Smith's leadership, the Housing and Food division of engineers cabled orders to the steamboat companies to equip their vessels with refrigeration plants. They built a large cold-storage plant at Colón and purchased refrigerator freight cars. Thus, they established a line of refrigeration from the markets of the United States to the commissary stations along the line of the canal. All the employees had the opportunity to obtain an abundant supply of wholesome food at reasonable prices.

A bakery was started at the canal when Stevens walked into the office of Frank Maltby, a division engineer. He sat

---

[43]Stevens, "The Panama Canal," 959; Du Val, *And the Mountains Will Move*, 193.

down, lighted a cigar, put his feet up on the desk, and said, "Maltby, what do you know about a bakery?"

"Not a damn thing."

"Well, you can find out, can't you?"

"Yes, sir."

"Well," he said, "I want to bake ten thousand loaves of bread a day."

"I suppose you mean yesterday."

"Yep."

"I am going to the States tomorrow," Maltby said.

Those remarks constituted the sole and total instructions on the project. Stevens was a man of few words.[44]

Maltby went to the United States, visited many bakeries, and learned what machinery and equipment were used. He decided what was best for use on the isthmus, and through the Panama Railroad he purchased them along with material for a building. Later, as the bakery was being erected, Stevens stopped by and asked Maltby how he was getting along. The latter thought it was a good time to get back at the chief engineer and replied, "All right," and nothing more.[45]

A short time afterwards, an earthquake occurred at Kingston, Jamaica. A relief ship was loaded at Colón to take supplies to the distressed city, and among them were ten thousand loaves of bread from the new bakery.

In 1907, there was a daily output of eight tons of ice and twenty-five thousand loaves of bread. There was a laundry for at least three thousand workers. There were baseball

---

[44]Frank B. Maltby, "In at the Start of Panama, part 1: My Introduction to a Tremendous Project," *Civil Engineering* 15, no. 6 (June 1945): 260–62.

[45]Frank B. Maltby, "In at the Start of Panama, part 2: Getting Under Way with a 'Great Chief,'" *Civil Engineering* 15, no. 7 (July 1945): 322–24.

teams, musical bands, YMCAs, and clubhouses for games and entertainment. The employees were contented with their working situations and sometimes gathered for social fun. Stevens often attended but did not mix well socially. He was not a person whom anyone would ever slap on the back. Very fond of his wife, Harriet, Stevens was lost without her outside of business hours. She was a gracious lady who had gone through the hardships with her husband, and was greatly loved and admired for her thoughtful kindness.

∗   ∗   ∗

A short time after Stevens arrived at the canal, the workers living at Corozal gave a welcome party for him. They invited important citizens, including Governor Magoon, Canal Commissioner Shonts, and Assistant Chief Engineer John Sullivan. Stevens was asked to give a few remarks. He could write and present the most convincing reports and tell entertaining stories, but he couldn't make a light and smooth informal talk. He began his remarks with, "I expected something of this kind and I told Shonts what to say and he has said it and left nothing for me to say." Then in a more serious tone he continued,

> I presume that I have had as much or more actual personal experience in manual labor than anyone here—surveys, hardships, railroad construction in all its details and operation. Why, I remember when I was the only white man working with a section gang in Southwest Texas at a dollar and ten cents a day.

His listeners looked at each other and then enthusiastically cheered him.[46]

The next speaker was the Irishman, John Sullivan. He

---

[46]Frank B. Maltby, "In at the Start of Panama, part 3: When a Division Engineer Was All Things to All Men," *Civil Engineering* 15, no. 8 (August 1945): 359–62.

said, "I never worked for the government before but I have worked for the Chief for several years. The old man has told that story so many times he believes it." Maltby told the story to Mrs. Stevens the next day. With a rather pathetic reminiscent smile, she said, "Well, there was more truth than poetry in it."[47]

＊　　＊　　＊

While preparation for sanitation, building, feeding, and housing of employees was going on, work on the canal construction was progressing. General charge of designing the dams, locks and spillway was given to Joseph Ripley. His staff worked out all plans of those structures as they were built, with few changes.

Actual construction of the canal was handled by three division engineers. F. B. Maltby's jurisdiction extended from Colón to and including all works at Gatun dam, spillway, and locks. He also was in charge of the construction and operation of the dredges, tugs, and marine shops. Work throughout the Culebra Cut was supervised by D. W. Bolich. His section extended to and included the locks at Pedro Miguel. He oversaw the excavation and disposal of all material from the cut, no matter to what point it was taken. Activity in the territory from Pedro Miguel locks to deep water in the Bay of Panama was under the leadership of William Gerig. His crew was responsible for conducting the vast amount of surveying and test borings at the southern end. Also, they had to dredge the outside channel plus maintain and operate the marine shops in their division.

The division of meteorology and river hydraulics was

---

[47]Frank B. Maltby, "In at the Start of Panama, part 4: I Change Chiefs and Take my Last Job on the Canal,: *Civil Engineering* 15, no. 9 (September 1945): 421–24.

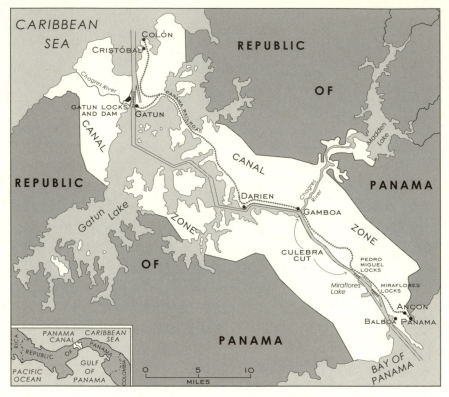

Map of the Panama Canal and Canal Zone.
Based on map in Encyclopedia Britannica.

headed by Ricardo Arango, a native of Panama. He was responsible for conducting all stream measurements and the extensive system of weather recording stations. John Stevens supervised the reconstruction of the Panama Railroad. The maintenance and operation of all machinery, including locomotives, steam shovels, and rolling stock, were in the charge of George Brooke.

In Washington, D.C., the canal commission had appointed David W. Ross to be manager of the purchasing and supply department. He was a very able manager and was both prompt and cooperative. All requisitions for supplies, materials, tools, and thousands of different items were speedily fulfilled by the purchasing department. The commission also hired another capable person, Edward J. Williams, to be the disbursing officer on the isthmus. His responsibilities were to handle the enormous amounts of money required for payments of salaries and wages, as well as to establish methods of timekeeping and identification leading up to the actual payments. He was a person who displayed unusual skill and honesty.[48]

* * *

In the fall of 1905, about four months after Stevens arrived to take over the building of the canal, Secretary of War Taft made a second visit to the Canal Zone. The first was in the previous year when John Wallace was the chief engineer. This time Taft wanted to clear up some questions with Panama, check the progress of the canal, and decide on questions of fortifications. He took with him General John P. Story, chief of artillery; Colonel Clarence R. Edwards, chief of insular bureau; Lieutenant Colonel Edward R. Black, Major George W. Goethals, and Lieutenant Mark Brooke.

There had been a question about the adequacy of the foundations for the Gatun locks, and Taft's party was assigned to inspect the site and report its recommendations. Under the supervision of Division Engineer Frank Maltby,

---

[48]Listing of the various departments and leader with their responsibilities were listed by John Stevens in Bennett, *History of the Panama Canal*, 212–17.

workers had bored hundreds of drill cores and dug a lot of test pits. The excavated material was loaded into large steel buckets that were hoisted to the surface with an old-fashioned windlass made of a trunk of a tree six or eight inches in diameter, supported across the pit, and fixed with a substantial crank at each end.

Maltby was assigned to meet Taft and the congressmen who had come to inspect the foundations. When they got off the train, Taft went over to the cool shade of a mango tree, sat down on a pile of railroad ties, and began to tell stories. Finally, he said, "Well, I suppose I have to look at what you have to show me."[49] Maltby asked Taft if he wanted to go down into one of the test pits so he could report that he had personally examined the foundations. One of the congressmen immediately said, "I think this is one time when we should insist on official precedence. The secretary should go first."[50] Taft was a large man, weighing nearly three hundred pounds. He took off his coat, stepped into the bucket, and was slowly lowered into the twenty-foot hole. At the bottom he inspected the foundation of the Gatun lock and then was winched to the surface. Engineer Maltby expressed his relief when they helped the secretary out, because next to the big man the rope on the windlass looked awfully small. Story and Edwards then took their turns at the inspection.[51]

During Taft's visit to the isthmus, a problem arose with one class of skilled laborers. They had a wage scale which, along with many granted privileges, resulted in more than 60 percent higher wages for similar work in the States, yet they threatened to strike if they did not get a higher salary

[49]Maltby, "In At the Start At Panama, part 3," 362.
[50]Ibid.   [51]Ibid.

increase, and some did stop working for a few days. Stevens refused to meet their demands. He told them that the ships were sailing north with few passengers, and that they could fill up the empty cabins if they didn't like their pay. Taft, who, next to the president, had supreme authority over canal affairs, supported Stevens' decision. The men went back to work, and the incident was closed. When the living and health conditions finally became what they should have been, the entire labor force settled into a business-like regimen. Affairs went smoothly, and much progress was made in the construction of the canal.[52]

*   *   *

Two legal situations in Panama were extremely bothersome for the chief engineer. One was the eight-hour work-day law forced on him by Congress. The other was the civil service statute. On August 1, 1892, Congress had approved a statute that stated:

> Service and employment of all laborers and mechanics employed by the government of the United States, or by any contractor or subcommittee upon any of the public works is limited and restricted to eight hours in any one calendar day. Violations will be $1,000, or six months' imprisonment or both.[53]

On June 30, 1906, Congress amended that law, which then read:

> An act relating to the limitation of the hours of daily service of laborers and mechanics employed upon the public works of the United States, approved August 1, 1892, shall not apply to alien laborers and to foremen and superintendents of such laborers

---

[52]John F. Stevens, "The Truth of History," in Bennett, *History of the Panama Canal*, 219.

[53]*Statues At Large of the United States*, vol. 27 (Government Printing Office, Washington, D.C., 1893), p. 340.

employed in the construction of the Isthmian Canal within the Canal Zone.[54]

That labor law caused a lot of consternation for Stevens as they tried to set up a work schedule for non-alien skilled employees and hundreds of alien laborers from Europe and the Caribbean Islands. It also was a hardship on the government and forced unit costs higher than intended.

A congressional act approved March 2, 1905, was a civil law that caused more anguish for canal employees:

> All laws affecting imports of articles, goods, wares, and merchandise and entry of persons into the United States from foreign countries shall apply to same articles, goods, wares, and merchandise and persons coming from the Canal Zone, Isthmus of Panama, and seeking entry into any State or Territory of the United States or District of Columbia.[55]

That law was a handicap in the selection of the labor force. Any laborer desiring to quit his job on the canal and go to the United States would be restricted on what personal property he could take in. Therefore, he was reluctant to take a job on the canal.

*   *   *

Toward the end of 1906, preparatory work for construction of the canal was almost complete. Under Stevens' supervision the preparation phase had moved rapidly. In June, Congress had decided to build the lock-type canal. Railroad tracks had been laid, and operation stations had been installed for dumping the spoil from Culebra Cut. The yardage dumped each day amounted to 327,000 cubic yards. Forty-eight steam shovels were at work on the line

---

[54]Ibid., vol. 34, pp. 33–34.   [55]Ibid., vol. 34, p. 843.

of the canal. Design work for locks and dams was under way. The working force was well-organized, well-housed, and well-fed, and morale was high. With this background Stevens predicted that the Panama Canal would be open for traffic by January 1, 1915. (His foresight was quite accurate, as the self-propelled crane boat *Alex LaValley* made the first passage through the canal on January 7, 1914. On August 3, 1914, the first ocean-going steamer, SS *Ancon*, sailed through, and on August 15, 1914, the canal was opened to commerce.)

As great as those achievements on the isthmus had been, there had developed in Washington a critical undercurrent. Secretary Taft did not want Chief Engineer Stevens and Isthmian Canal Commission Chairman Shonts to dominate relations with the Panamanian government. He thought they would take the same attitude toward labor on the isthmus that they would if working on some railroad, which was the background of both men.[56] He assumed that Stevens and Shonts would ignore government labor laws, insist on long hours of work, and prevent strikes or other means of obtaining better working conditions. When Stevens learned of Taft's assumption, he realized that, along with some of the labor problems he had experienced and hindrance from federal labor legislation imposed on Panama, now the serious question was raised of what method to be used to construct the canal: Should the work be let to an independent contractor or built by day labor under direct supervision of the canal commission?

Stevens knew that large railroads of the United States, when building the transcontinental lines, used the contract method. Contractors had the following of experienced

---

[56]Du Val, *And the Mountains Will Move*, 226.

employees, foremen, superintendents, and engineers. Being a railroad man also, Shonts agreed with Stevens' contract proposal. Stevens had both built railroads with contractors and worked for them. Even though he favored them, he was also practical and realistic. He wanted the bidding to be based on "intelligent specifications."

With preparatory work rapidly nearing completion, a decision as to the method of construction became urgent. Was it to be let by contract or built by the government? He decided to recommend placing the actual physical parts of construction under the contract system. He hoped the government could attract the cooperation of strong railroad and general contractors who could combine their strength and influence in forming a strong syndicate that would gain the respect of the business world. They could become an organization of the best construction and material experts possible. He presented his thinking to Chairman Shonts and stressed the percentage profit system under which both government and contractor would be protected. There would be a deduction in the contractor's costs for each month he went over agreed estimates, and a bonus for each month costs were saved. Included was the assertion that the Isthmian Canal Commission, through its chief engineer, would be the sole authority and arbitrator.

Shonts presented this proposal to the commission which requested that Stevens submit a form of contract for its consideration. Because of the multiplicity of details involved, it was impossible to ask for bids upon separate unit quantities, so Stevens used as a basis a modified form of percentage contract that he had successfully employed in heavy railroad construction work. He prepared a form, modified in such a manner as to make the chief engineer supreme in all matters and questions that might arise, chances for which were numberless.[57]

The commission approved Stevens' percentage plan and sent it to Roosevelt, who asked Stevens to come to Washington and discuss it. He arrived at the capital in May 1906. After listening to Stevens, the president took the proposal to his Cabinet. Both the members and the president gave the plan tentative approval, except Elihu Root, secretary of state. He insisted on injecting technical conditions, which in Stevens' opinion weakened the plan's value because it would make the percentage proposal unattractive to the class of contractors he desired to interest. However, Root's views prevailed.

In explaining his percentage contract to Congress, Stevens emphasized the importance of the government retaining control of sanitation and quarters in the Canal Zone. He suggested dividing the work into several parts instead of turning it over to one contractor. He knew that contractors were not familiar with work on the isthmus and did not have the proper equipment. He stated that those factors would require them to include all classes of contingencies in their estimates and cause delays in assembling the plant and beginning the work. Following the presentations of his contract system in Washington, Stevens returned to the isthmus.

The president was attracted by Stevens' idea of rapid construction and completion by contract. The next month he called Stevens again to the White House and told him he was "extremely anxious" (Roosevelt's expression) to have the canal under contract construction by November 1906. Back on the isthmus, Stevens consulted with another engineer

---

[57]Stevens, "The Truth of History," in Bennett, *History of the Panama Canal*, 221.

well-experienced in letting contracts. In July he wrote Shonts a warning that he, Stevens, was strongly opposed to hiring contractors by advertisement and letting the entire work to any one firm. That would take too much power from the commission and make possible combinations to control prices. This could result in disasters, cause delays, and postpone the completion of the canal. He emphasized again the need for keeping control in their hands and under the directions of the commission engineers.

He enclosed in the letter a process for hiring a group of ten or fifteen first-rate railroad and general contractors. Each firm to be selected needed to have a capital of $25 million and be able to give a $10 million bond for construction of the canal. The work would include dredging of the sea-level portions; construction of locks, dams, and regulating works; excavation from Gatun to the Pacific locks including Culebra Cut; and relocating the Panama Railroad and breakwaters. Agreement was to be on a percentage-profit basis, with deduction for overrunning agreed estimates, and bonuses for each month saved. The commission would retain control of government and sanitation, material and supplies, municipal engineering and waterworks, building construction, labor and quarters, and repair of machinery. Stevens also wrote to Roosevelt mentioning he had sent a letter to Shonts and expressing hope to have the matter decided and to have construction begun by mid-October.[58]

Shonts and the Isthmian Canal Commission endorsed Stevens' proposed process and forwarded it to Roosevelt and Taft in August for their approval. Taft rejected the plan because he thought that hiring contractors should be by competitive bidding and that Stevens' plan of selecting them

[58]Ibid.

might have political repercussions. He told that to Shonts, who, not being experienced with letting contracts and not feeling as strongly as Stevens against advertising for bids, acquiesced to Taft's desires.

The commission chairman sent Stevens a message that he and Richard Rogers, another member of the commission, would revise Stevens' plan and send him a copy.[59] Roosevelt's response to the proposal was to consult with Shonts, Cabinet members, and a few contractors. Shonts told him he agreed that the canal should be built by contractors, but objected to the commission and the chief engineer being the final arbitrators in selecting them. His opinion began a strong disagreement between him and Stevens. The president agreed with Taft that choosing contractors without competitive bidding would arouse unpleasant political reactions.

On October 9, 1906, Roosevelt had the commission issue a public invitation for contractors' bids. Writing up the bids was difficult and inadequate. They called for excessive financial qualifications from the bidder. The result was a meager response.[60]

President Roosevelt desired to see the progress of work that had been going on at the canal. In the summer of 1906, he announced his intention to visit Panama in November. His reason for going that month was that it would be the height of the rainy season, and he wanted to see how the employees handled the worst of working conditions. He left on November 9, 1906, for a three-day inspection of the Canal Zone. It was the first time an American president had left the country while still in office. He sailed on the

---

[59]Du Val, *And the Mountains Will Move*, 250–51.
[60]Ibid.

new battleship *Louisiana*, escorted by two cruisers. His wife, Edith, accompanied him.[61] They arrived at Panama on November 14. A large welcoming program had been planned for him at 8 A.M. the following day. A small cannon had been secured from the Panama Army to fire a presidential twenty-one-gun salute and school children were to sing "The Star-Spangled Banner." At 7 A.M. a naval launch arrived with the president and his party, an hour ahead of time, and there was no one on hand to greet or salute him. Roosevelt climbed on the dock and began talking to carpenter employees. A messenger was quickly sent to Washington Hotel a mile away where the official welcoming party, including Secret Service men, were having breakfast. They all scurried to the dock. Troops from the Panamanian Army arrived and fired the gun salute. The school children were brought and lined up with their backs to the president. (They sang their song that way because the president wasn't where he was supposed to be.)[62]

One day at the Hotel Tivoli in Ancón, Roosevelt excused himself from his hosts, newspaper people, and others, saying that he would retire to his room. Instead, he rushed to the Ancón Hospital wards and began talking to the patients about their treatment and care. Colonel Gorgas, the medical officer, was prepared to show him the hospital and everything he wanted to see if the president had asked him. Engineer Maltby's opinion was that the president seemed obsessed with the idea that someone was trying to hide something from him.[63]

The president spent another day being photographed on

---

[61]McCullough, *The Path Between the Seas*, 493.

[62] Maltby, "In At the Start At Panama, part 4" 421.

[63]Ibid., 421–22.

a steam shovel holding the levers and talking to a group of men. That evening had been set aside for a meeting with division engineers and heads of departments. Stevens had instructed them to come prepared with information, as the president may want to ask a lot of questions. They sat in the hotel lobby until after 11 P.M., while Roosevelt talked to some steam-shovel operators who were spending the evening socializing at the hotel. It became so late, and the president was so tired, that there was only opportunity for the division leaders to be introduced, shake hands, and retire, still in possession of all the information they had brought.[64]

On the last day of Roosevelt's visit, Stevens told his principal chief engineer, Maltby, that he and Chairman Shonts were both worn out. Maltby would have to show the president what he wanted to see. Most of the day was spent with the president giving speeches to various groups. That evening a reception was given on the dock at Cristóbal for President and Mrs. Roosevelt, and they sailed for home. Later that evening as Stevens and Maltby were visiting, the chief engineer said to his assistant, "Maltby, I know you pretty well now and without raising the question of your competence, if you were chief engineer you wouldn't last thirty minutes." "No," Maltby said, "I wouldn't want it if I could get it." "Well then," replied Stevens, "why should I stay here?"[65] Either the chief engineer was extremely tired and had had a bad day, or was beginning to entertain thoughts of resigning.

President Roosevelt's trip to Panama seemed to have buoyed his spirit and given him more confidence for the ultimate success for the construction of the canal. In a letter to his son, Kermit, written aboard ship, homeward bound, he wrote, "Our visit to Panama was most success-

---

[64]Ibid., 422.   [65]Ibid.

ful as well as most interesting. . . . Stevens and his men are changing the face of the continent, are doing the greatest engineering feat of the ages, and the effect of their work will be felt while our civilization lasts."[66] In his report to Congress on December 17, he expressed utmost confidence in Stevens and his staff.

While the planners in Washington fumbled with contracts and contractors, activities on the isthmus kept progressing. Equipment continued to arrive and was assembled and placed at work. Excavation yardage steadily increased throughout the winter on the Culebra Cut, and it was expected to increase once the dry season arrived.

One of the last incidents of 1906 was a flood of the Chagres River. At Gamboa the river reached a height of 79.9 feet above sea level, which was about 40 feet above the bottom planned for Culebra Cut. The flood showed that a dike would have to be placed across the cut at Gamboa to protect it from the flood waters of the Chagres. Also, the old French diversions on each side would have to be placed in service to protect the canal work from the turbulent river.[67]

During the long wait for construction bids to be opened on January 12, 1907, progress on the canal was so rapid that Stevens was praised for accomplishing so much without contractors. The need for contract work was disappearing. When the bids were opened, there were only four from which to choose: George Pierce and Company, Frankfort, Maine, 7.19 percent of profits; William J. Oliver and Anson M. Bangs, New York City, 6.75 percent; MacArthur-Gille-

---

[66]Elting E. Morison, ed., *The Letters of Theodore Roosevelt*, vol. 5 (Cambridge, Mass.: Harvard University Press, 1952), letter 4140 to Kermit Roosevelt, 496; letter 4282 to Richard Rogers Bowker, p. 629, Alfred D. Chandler, Jr., "Theodore Roosevelt and the Panama Canal: A Study in Administration," Appendix I, pp. 1547–54 (hereafter cited as Chandler, "Theodore Roosevelt").

[67]Du Val, *And the Mountains Will Move*, 247.

spie Company, Chicago, 12.50 percent; and North American Dredging Company, San Francisco, 28 percent. Commission Chairman Shonts cabled Stevens asking advice for the best thing to do. Stevens' reply was that the lowest bid was too high, and he wanted to know the record and capacity of each firm before deciding.[68]

In the midst of this decision-making, Theodore Shonts received an offer to lead a transportation merger in New York City. He decided to take the job and submitted his resignation as chairman of the Isthmian Canal Commission to President Roosevelt. On January 22, 1907, the president accepted Shonts' resignation, to take place on March 4. His resignation was a personal regret to his many friends. Some reported he had become disgusted with Washington red tape and political interference. Others felt it was due to relations with Stevens. The chief engineer's knowledge and ability were noticeably superior to those of Shonts, which caused some of the disharmony between them.[69] After receiving Shonts' resignation, the president appointed Stevens as chairman of the commission.

Prior to leaving his post and after contractors' bids were received, Shonts sent his agreement to Roosevelt and Taft to hire contractors Oliver and Bangs. He had earlier promised Stevens he wouldn't express his decision until he consulted with him first. They agreed to award the contract to Oliver and Bangs, who had bid at 6.75 percent. They also permitted Oliver to have ten extra days to form a $5-million corporation with new associates. Bangs was eliminated from his part of the bid when Oliver obtained new associates. When Stevens learned of their decision, he cabled Shonts that he believed it would be a mistake to award the

---

[68]Ibid., 252.   [69]Ibid., 252–53.

contract to Oliver, whom Stevens considered not qualified by "nature, experience, or achievement." He objected also because the changing of associates was the same as allowing a new bid. In addition, Oliver had caused much ill-will among the canal's best employees by his published interviews.

Stevens also cabled to Secretary Taft his same objections, and Taft requested more information regarding his disapproval. Stevens replied that the main object of the contract was to gather large numbers of the best specialists for each class of work, and that ability and fitness were worth more than money. The bid under consideration was simply a proposition by one man with not much ability. Oliver's published interviews—with remarks that he planned to use Southern convicts as laborers—had created hard feelings on the isthmus and would produce situations that could not be controlled. The chief engineer went on to state that as a business proposition, the contract should never have been advertised. He ended by asking Taft to take time to make the best arrangements and not to hire Oliver.[70]

Again, Taft asked Stevens for more enlightenment on Oliver's published interviews. In a lengthy cable Stevens explained that nearly every paper from the States contained reports of interviews which stated that the contractor's crew would go down to Panama at once with large numbers of steam-shovel men to make the dirt fly; that the workers would be properly housed and fed; and that Oliver favored bringing thousands of Negro convicts from the South to serve as laborers. Stevens added his observation that "a Napoleon is not needed here, but (rather) such an organization as outlined in my letter of July 27 to Chairman

[70]Ibid., 254–55.

Shonts," which he asked the secretary to look up. Taft was persistent and cabled back that Oliver had stated the reported press interviews were unfounded. However, the secretary held Oliver's bid in abeyance.[71]

On February 8, 1907, President Roosevelt sent Stevens correspondence in which he expressed his astonishment that Stevens should think the contract matter was wrong, especially since he had made the plan in the beginning. He asked for Stevens' assistance in carrying the policy through to the finish. Stevens' answer the next day stated it was the first advice he had received on the contract matter. Previous information had indicated to him that it was simply a one-man project by one contractor. He then listed the points differing from his original plan: the capital required was one-fifth; the bond required was one-half; current payment was to be on percentages; and the contractor was to supply all labor. He reminded Roosevelt that he had objected to the last two provisions at a meeting in December, but was overruled. Then he added that Roosevelt would receive a personal letter which might clarify the situation.[72]

On January 30, Stevens wrote a letter to President Roosevelt that arrived at the White House on February 12, 1907. It was six pages long and written in a much different spirit than he had displayed in the past. It contained a sense of bitterness and exhaustion. He expressed appreciation for the president's approval of his work. He stated that when he accepted his job he thought the canal work would be "a purely business proposition." Instead of being permitted to concentrate on building the canal, he had been forced to "fight a continuous battle with enemies in the rear" and had been "continually subject to attack by a lot of people, and

[71]Ibid., 256–57.   [72]Ibid., 257–58.

they are not all in private life, that I would not wipe my
boots on in the United States." He complained about the
responsibility and strain put upon him. In closing, he
asked the president to relieve him in two or three months,
if it did not "embarrass in any way your plans."[73]

Roosevelt was shocked by the letter. Although the letter did not contain a formal resignation, he perceived that
the chief engineer seemed to be breaking under the strain
and pressures. He immediately sent the letter to Secretary
Taft with a note stating, "Stevens must get out at once."
He indicated that even if Stevens should change his mind
as he had before, he would not reconsider the matter
because of the "tone of the letter."[74]

On February 14, Roosevelt wrote Stevens to inform him
that his resignation was accepted. He would be relieved as
early as possible, probably by an Army engineer. The truc-
ulent tone of Stevens' letter was not only annoying to the
president, it demonstrated that Stevens was not adapting
himself to the pressures of public work. Also, he was
threatening to quit his job at a critical moment in the canal's
history. Theodore Shonts had just resigned, and it was just
before the decision on contracts was to be made. Roosevelt
considered the canal his greatest work, and it could not be
built with chief engineers leaving every year like Wallace
and Stevens had done. Days of preparation and organization were over. Construction progress was such as to permit its being placed in permanent hands that would see it
through to the end. He wrote, "I propose now to put it in
the charge of men who will stay on the job 'til I get tired

---

[73]Chandler, "Theodore Roosevelt," 1553; Du Val, *And the Mountains Will Move*,
238–39.
[74]Ibid., 259.

of having them there, or 'til I say they may abandon it. I shall turn it over to the Army."[75]

Roosevelt discussed his decision with Secretary Taft and both agreed that the change had been necessary. The president asked for a recommendation of someone from the Corps of Engineers to be chief engineer, and Taft consulted with General Alexander Mackenzie, chief of engineers, for a capable man in the Army Corps of Engineers to replace Stevens as chief engineer on the canal. Mackenzie recommended Major George W. Goethals. On February 18, 1907, the president sent a message to Major Goethals to meet him at the White House. Goethals and his wife were entertaining dinner guests, but the major changed into uniform and went right away. Roosevelt told him of the events that had made a change necessary at Panama. He expressed regret at Stevens' resignation and explained that he wanted to secure continuity of leadership. He could not afford to have his principal administrators leave or threaten to leave in time of trouble. He told the major he believed that army officers were more accustomed than civilian administrators to the publicity and the irritations involved in government work. He concluded that since Shonts and Stevens had solved the transportation problems of the construction, army engineers, with their training in the building of locks and dams, were technically able to carry out the work in Panama. The president then assigned Major Goethals to be the chief engineer of the Panama Canal.[76]

Whatever caused the sudden change in Stevens' attitude remained a mystery. Various persons have attempted to explain it, saying that he got dissatisfied with negative press

---

[75]Ibid.

[76]Ibid., 260–61; Chandler, "Theodore Roosevelt," 1553.

releases. Some blame it on government red tape. Some are sure it was because of his emotional opposition to contractors' bidding. Others cite insufficient administrative powers. Still others claim it was his overzealous choice of civilian and railroad engineers over Army engineers. Regardless of what critics have expressed, Stevens stated it was purely for personal reasons. Speaking at the annual meeting of the American Society of Civil Engineers in Denver on July 13, 1927, he said:

> An erroneous idea long prevailed that the work of those two years [as chief engineer] covered only that of preparation—far from such was the case. While the preparatory jobs were being pushed, so also was the actual construction of the canal proper. The dredging of the ocean approaches, the excavating for the locks at Gatun and Pedro Miguel, the construction of the dams at both places, was proceeding.
>
> . . . In early 1907, material removed from the Culebra Cut averaged about 750,000 cubic yards per month. The whole elaborate machine which had been built up so quickly and carefully, was working smoothly, and the output of all classes of construction was increasing daily at a rate that exceeded even the sanguine expectations of the chief engineer.
>
> . . . Knowing full well that there had been carried out—even beyond the letter—the promise made to the president (that I would not agree to remain until the completion of the canal, but would remain until the failure or success of the undertaking was assured according to my own judgment), and knowing that the latter was alive to the fact, I resigned from the service of the canal. Many reasons, mostly from irresponsible scribblers, were given as the cause of my resignation. . . . Not the slightest friction ever developed between the chief engineer and the president, or, in fact with anyone high or low, connected in any manner with the enterprise. The reason for the resignation was purely personal, and involved nothing in regard to the work nor with my parties associated in it. . . . I have never declared these reasons, and probably never will as they were private and of no particular interest

to the public. . . . I felt well-assured that the work which had been so near to my heart had been given into competent hands, as the future proved in every way to be the case.[77]

Theodore Roosevelt never mentioned Stevens again, even in his autobiography, where he wrote about building the Panama Canal. As a result, few publicly acknowledged Stevens' contributions to the engineering feat. However, the engineering community would later give Stevens its highest honors.

While his relationship with Roosevelt was never repaired, Stevens always held him in high regard. During an illness that would end his life, Stevens said to his son, John F. Jr., "Son, the next time you come I shall not be here. On the mantel are the pictures of the only two men who ever influenced my life and I wish you to have them." The pictures were of Hill and Roosevelt.[78]

The same week that he selected Goethals, Roosevelt conferred with the firm of contractors regarding their bid. By then he had become convinced that Stevens had been right about contract bidding. Only by negotiating a contract could the government get the best contractors at a price equitable to both government and contractors. Roosevelt refused their bids. Thanks to Stevens' brilliant work in preparing the physical plants and organizing the labor force on the isthmus, there was no longer a real need for the specialized skills of expert contractors. On February 26, 1907, the president announced that the bids of the private contractors had been rejected, and that Army officers would take charge of the canal construction. With that announcement his major administrative concerns in Panama were over. Goethals and

---

[77]Stevens, "The Panama Canal," pp. 960–61.
[78]"The Engineering Genius History Forgot," American Society of Civil Engineers News Release, internet.

his Army engineers took over the administrative duties and completed the canal in 1914. There was much criticism on the isthmus as people tried to assess responsibilities for taking Stevens away. They blamed party politics, an incompetent commission, a vacillating Congress, and transcontinental railroads. The local paper, *Panama Stars and Stripes*, March 22, 1907, expressed the common view of the local citizens: "We here on the ground, and every American worker on the isthmus, will say that the credit very justly belongs to John F. Stevens."[79]

In their communication, Stevens and Goethals agreed they would be ready to make the transfer of authority on April 1, 1907, and notified Secretary Taft of this. On March 27, Stevens, who was then both the chief engineer and chairman of the Isthmian Canal Commission, called a meeting of the commission. He submitted his resignation, effective April 1, and confirmed Goethals' appointment as chief engineer, to become effective on the same date. It was Stevens' only meeting with the commission as its chairman. On March 2, Major Goethals was appointed to the rank of lieutenant colonel.

In early March, Lt. Colonel Goethals arrived in Panama accompanied by Major David Gaillard, a new member of the Isthmian Canal Commission recently reorganized by Roosevelt. Their reception was cool. Stevens and Gorgas met them at the pier where they landed. No formal welcoming program had been planned. No arrangements had been made for a place for Goethals and Gaillard to stay. They were finally moved into the quarters of Dr. Gorgas and had little privacy.[80] In a short time Lt. Colonel Goethals

[79]Du Val, *And the Mountains Will Move*, 269.
[80]McCullough, *The Path Between the Seas*, 532.

observed that the local newspaper openly deplored the situation of military rule. He also realized Stevens' railroad employees had little respect for Army engineers. It was rumored that Army men had only technical training and they had never made a success "as executive heads of great enterprises." He later wrote, "It was predicted that if they ... were placed in charge of actual construction the canal project was doomed to failure."[81]

Several days after the newly-appointed chief engineer arrived, a welcoming reception was held for him. Stevens didn't attend. Lt. Colonel Goethals sat at the head table and listened while the toastmaster extolled Stevens and made unkind remarks about the military. Goethals was angry. It was an evening he would never forget.[82] When he stood to give a message, he said:

> Mr. Stevens has perfected an organization which, if maintained, will carry this canal through to its completion. I want to say here that it's my intention to keep that organization as he has established it. I have understood that there was some little feeling on account of militarism, but I want to state here that I do not expect a salute from any man on the job. I am no longer a commander in the United States Army. I now consider that I am commanding the Army of Panama, and that the enemy we are going to combat is the Culebra Cut, and the locks and dams at both ends of the canal.[83]

On March 30, 1907, Secretary of War Taft arrived in Panama with some consulting engineers and a group of congressmen. Stevens met them and accompanied them to Panama City. On the way Taft viewed Gatun and visited Culebra with Stevens, who gave detailed explanations of

---

[81]Ibid.   [82]Ibid., 532–33.
[83]Du Val, *And the Mountains Will Move*, 268.

the work. At Culebra Stevens introduced D. W. Bolich, the division engineer of that section, proudly proclaiming him as "the man who is digging the Culebra Cut." At midnight on Saturday, March 31, John F. Stevens terminated his services with the canal. Upon the shoulders of Colonel George Washington Goethals fell the heavy responsibility of constructing the Panama Canal.[84]

The next day, April 1, Stevens again went with Taft's company on their inspection tour. Since Goethals was then the chief engineer, Stevens held back to let him explain the work, but Goethals insisted that Stevens do the talking.

Stevens prepared to leave the isthmus on April 7. The employees planned a farewell reception for him to be held at Cristóbal the night before. Special trains came, crowded with guests from all along the line. Many planned to remain overnight to see him depart the next day. Goethals did not attend the party. At 9:30 P.M. Stevens arrived, accompanied by close friends. The Isthmian Canal Commission band, which Stevens had sponsored, played "The Conquering Hero Comes." The crowd parted to make an aisle to the podium for him. General Superintendent W. G. Bierd was spokesman for the employees. He said it was unnecessary to explain the reasons for the demonstration of such regard and affection, but he did want Stevens to know the full measure of esteem in which he was held. Acknowledging that Stevens was no easy taskmaster, Bierd stated that he had won their respect and affection because they were convinced he was a man always able to overcome the problems involved and whose decisions always impressed his subordinates with the belief that they had been made for reasons he believed to be right.

After Bierd's remarks, Stevens addressed the crowd. He

---

[84]Ibid., 269–70.

gave credit to his predecessor, John F. Wallace, for the organization he had inherited and that he had modified. "Mr. Wallace," he said, "is one of the ablest engineers in our country, and his work here is what I would have expected of him ... earnest, clever, and thorough."[85] He also gave recognition to Col. Gorgas. He said that two years before he (Stevens) had been almost as overwhelmed as the president was by the vast volume of preparatory work required. "Until Colonel Gorgas lifted the dark cloud which unsanitary conditions placed over the work, I was doubtful of success." When that cloud was lifted he knew the men with him would complete the canal. He then appealed to the workers gathered to take their little differences to Colonel Goethals, not to Washington. He asked them to show the same loyalty to Goethals as they had to him.[86]

Following Stevens' message, Bierd presented to him on behalf of the employees the copy of a petition, signed by ten thousand individuals, requesting him to reconsider his resignation and remain in Panama. In addition, they gave him a gold watch, a diamond ring, and a set of silver table flatware with an engraving that depicted the completed canal.[87] Stevens was visibly moved by the demonstration and realized it marked the end of an episode in his life. With its passing went forever his high hope of becoming the builder of the Panama Canal. It was almost midnight when he left the farewell party and boarded a vessel in the bay to spend the night.

The next day, Sunday, April 7, 1907, he went aboard the SS *Panama*, which waited at the dock. Hours before its sail-

---

[85] Carr, "The Panama Canal," p. 268.

[86] Flood, *Congressional Record*, House, 9288; and Inkster, "John Frank Stevens, American Engineer," 85.

[87] Flood, *Congressional Record*, House, 9288.

The SS *Cristobal* in Culebra Cut, Panama Canal, at Paraiso,
looking north from Cerro Luisa, August 4, 1917.

ing time, the largest crowd since the United States occupation, including many who had stayed in Cristobal overnight,
gathered at the pier to see Stevens sail away. At noon the
Panama left her dock and headed toward the Caribbean Sea
during the cheers of the crowd and the whistles of the ships
in the bay. The band played "Auld Lang Syne." Stevens stood
at the ship's rail with his son, looking pale and sad.[88]
    The Panama Canal was completed under the supervi-
sion of Colonel George Goethals in August 1914. Its length

---

[88]Ibid.

from deep-water at Cristobal breakwater on the Atlantic side to deep-water at San Jose Rock at the Pacific terminus is 51.2 miles, and 40 miles from shoreline to shoreline. Its axis is northwest-southeast, the Atlantic end being 27 miles west of the Pacific end. The area of Gatun Lake is 163.5 square miles.

During the years following his work on the Panama Canal, Stevens received many verbal comments and letters complimenting his service. One that he particularly appreciated was from a fellow engineer:

> Dear Sir: Fearing I may not see you before we sail, I want to write a word of appreciation of your splendid work here on the isthmus. I had not the faintest conception of it before I came here.
>
> I can imagine something of the chaos that existed when you came here but the order which you have brought out of the confusion is marvelous. Your organization is most complete. You have proven yourself not only an engineer of the highest rank, but an able and thorough executive. The world will some day realize the debt of gratitude it owes to John F. Stevens. With best wishes for your future, I am,
>
> Yours very truly,
Sereno E. Payne[89]

Although at times feeling overshadowed by Stevens, his successor, Colonel Goethals remained an open admirer of him. After his first five days of inspection following his becoming the chief engineer, Goethals remarked:

> Mr. Stevens has done an immense amount of work . . . has perfected such an organization . . . there is nothing left for us to do but to just have the organization continue the good work it has done and is doing. As I go over the line and see what he has accomplished and the organization he has perfected, I cannot see why he has resigned. . . . Mr. Stevens has done an amount of work for

---

[89]Stevens, "The Truth of History," in Bennett, *History of the Panama Canal*, 224.

which he will never get any credit, or, if he gets any, will not get enough.[90]

William H. Galvani, writing for the *Oregon Historical Quarterly*, describes a meeting with General Goethals (promoted from lieutenant colonel) in 1922. On February 27, the engineering profession of Portland held a reception for General Goethals. Before the meeting Galvani and Goethals met and discovered they had known each other in the Army. While they engaged in reminiscing, members would approach them now and then with visitors whom they would introduce to the general, always remarking that he was the "Genius of the Panama Canal." After this had occurred several times, Gen. Goethals said to Galvani:

> Do you know there is nothing more annoying to me as the statement so generally made in my presence that I am the "Genius of the Panama Canal"—I do not like it. Frankly, it is a fact, that Mr. Stevens devised, designed, and made provisions for practically every contingency connected with the construction and subsequent operation of that stupendous project, and when he turned over the office of chief engineer to me, everything was in the very best working order. . . . It is therefore to him, much more than to me, that justly belongs the honor of being the actual "Genius of the Panama Canal."[91]

On March 23, 1925, the American Society of Civil Engineers presented their highest award, the John Fritz Medal, to John Stevens. During his introductory remarks, John R. Freeman, chairman of the arrangements committee, said:

> I should say a word about the recipient of the medal . . . for I have known him more than twenty years, and for the past eight-

---

[90]William H. Galvani, "Recollections of J.F. Stevens and J.H. Mitchell," *Oregon Historical Quarterly*, vol. 44 (September 1943): 317, citing Joseph Bucklin Bishop and Farnham Bishop, *Genius of the Panama Canal*, 140.
[91]Ibid., 314.

een years, knowing how he brought order into chaos and organized a peaceful army of 30,000, and planned the work as no one else in the world could have planned it and set all going efficiently, I have regarded him as the real builder of the Panama Canal.

I happened to be at Panama when the transfer from the civil to the military authorities took place and could but feel the sorrow that was in the air at having Stevens leave the grand corps that he had been leading. I recall the statement made to me at that time by Colonel Goethals about the esteem and affection in which Mr. Stevens was held by every one on the Isthmus. He said, "I never have seen so much affection displayed for any man, and if I can so carry things on as to build up a similar feeling when I get through, it will be the proudest work of my life."[92]

When General Goethals died in January 1928, the *New York Evening World* paper printed:

General Goethals never boasted of his great accomplishments, and when the canal was mentioned in his presence he always insisted that two men, Theodore Roosevelt and John F. Stevens, had far more to do with the successful building of the canal than he. . . . "Stevens," he would say in his quiet way, "was one of the greatest engineers that ever lived, and the Panama Canal is his monument."[93]

In 1962, a major traffic circle in Balboa on the Canal Zone was designated as Stevens Circle and a cenotaph, engraved with the words "The Canal Is His Monument" placed there.

In the Senate on April 25, 1956, Senator Frederick Payne of Maine noted that year to be the 50th anniversary of the decision of the Congress to approve the lake-lock plan for building the Panama Canal. That same day was also the 103rd anniversary of the birth of Stevens. The Senate voted to approve Payne's request to have a statement and poem

---

[92]John Fritz Board of Award, "Presentation of the John Fritz Gold Medal to John Frank Stevens," 18–30.

[93]"An Engineer's Recollections," 52.

about Stevens printed in the *Congressional Record.* The poem
is the work of Maurice H. Thatcher, a former civil governor of the Canal Zone, a member of the Isthmian Canal
Commission, and a member of Congress.

JOHN F. STEVENS: A TRIBUTE
Amongst all those whose labors cleft the land
To blend as one, the seas of Panama
There was none greater than John Stevens; and
The passing years bear witness. He foresaw
More clearly than the others had foreseen
The value of the plan for lock and lake,
And led Authority in doubt between
Diverse designs—the wiser choice to make.
Possessed of genius rare, with skills supreme
And ripen'd knowledge gained from ventures vast
He shaped the molds to vitalize the Dream
Which had so long persisted in the past,
His all he gave to serve the Isthmian Task:
What more could men demand, or duty ask?[94]

---

[94] *Congressional Record*, Senate, Sen. Frederick G. Payne speaking on the anniversary
of the birth of John F. Stevens, 84th Cong., 2nd sess. (May 29, 1956), pt. 7, 9285–89.

5

A Return to the Railroads

After Stevens severed his connection with the Panama Canal, he and his wife intended to visit Europe. He had endured two years of exacting work on the isthmus and felt the need for rest and a complete change of environment. However, when the ship carrying him from Panama arrived in New York, a representative from the New York, New Haven & Hartford Railroad came aboard before Stevens disembarked. He told the engineer that the president of the railroad, Charles H. Mellen, wanted to confer with him before Stevens made any decision about his future. Stevens had known the president a long time, so he agreed to his request. A few days later they met at the Hotel Touraine in Boston. The two friends renewed acquaintance, and Mellen told Stevens that the New York, New Haven & Hartford Railroad was planning to take control of the Boston & Maine line, an independent railroad originally built by the New York & Erie Railroad Company. The merger would involve changing presidents, and he asked Stevens if he would accept that position on the Boston & Maine when the merger occurred.

Stevens hesitated to make the commitment because of the European trip he and his wife had planned. He discussed it with Mrs. Stevens and they agreed that since they both were originally New England people and had many relatives in that section, they would live there and forego their trip. Stevens consented to accept the generous offer. In a short time it appeared there would be a delay in the plans for merger due to financial problems on the New Haven. That company asked Stevens to come to their offices and organize a group to evaluate all of the New Haven's steam railroad properties. They assured him the delay would likely be a short one. Stevens agreed, but the work was more difficult than he had expected. It took a large force of engineers to complete it, as they had to count, classify, and list every item of property involved in steam engine transportation. (Some of New Haven's line was electrified and not included in the survey.) After that, they had to put a proper value on each piece based on the cost it would take to replace it. The evaluation was determined by the length of time it had been in service with allowance for depreciation. That project was the first attempt of such evaluation by any of the United States railroads.

The result showed that New Haven's capital stock was adequate. Its financial troubles came from unwise investments in outside properties at excessive cost in the New Haven railway's efforts to monopolize all transportation in its area. In addition, it had spent large amounts of borrowed funds for improvements that did not produce additional revenue nor reduce the expenses of operation.

Following Stevens' successful evaluation of their railway, the New Haven officials asked him to be vice president, and he accepted the position. His duties would be to remodel and modernize the operation system. In a short time he realized he would be unable to improve the financial situation, so he concentrated on improving methods of operation and getting more cooperation from the employees.

Labor unions and company officials had unknowingly prevented a loyalty and spirit of togetherness among the workers. One evening he rode in the cab of a locomotive pulling a train from New Haven to Boston. By the end of the journey the engineer had recognized Stevens as the railroad vice president and said, "I appreciate very much your riding with me. I have been running an engine on this road for nineteen years, and you are the first executive who ever stepped into my engine, or I believe, who ever spoke to me."[1]

W. G. Bierd, who had recently been manager of the Panama Railroad and had resigned and returned to the States, was hired as general superintendent by the New York, New Haven & Hartford Railroad. Stevens was pleased to work with him again. Bierd's knowledge of details and modern methods of operation produced a number of changes that improved the railroad.

Long before Stevens left the New Haven, he knew that the rosy outlook that had been told to him when he was first hired was only a fanciful dream. However, two important improvements were made on the New Haven under Stevens' tenure. One was the electrification of the railway from Woodlawn to Stamford, New York, with a high-voltage alternating current (AC) system. The other was the equipping and putting into use a large modern engine repair shop built at Readville, Massachusetts. Locomotive service, which had deteriorated due to lack of upkeep, was increased by this action.

In 1909 his former employer, James J. Hill, approached him in New York. Hill expressed a desire to build a rail-

[1]"An Engineer's Recollection," 54.

road south through Central Oregon and asked Stevens if it could be done and if he would do it. Stevens replied it could be done, and he would do it, although at that time neither one knew just how it could be done.[2] Hill accorded Stevens the position of president of the Spokane, Portland & Seattle Railway, and Stevens accepted it. He then went to Charles Mellen and tendered his resignation as vice president of the New York, New Haven & Hartford Railroad. Mr. Mellen was resentful at first, but, following a discussion with Stevens, he accepted his resignation.

As Stevens resumed his railroad career with Hill, he probably didn't anticipate the intrigue—one of the most bitter conflicts in competitive railway construction and the multiple court decisions—in which he would be involved. For years James Hill of the Great Northern lines and Edward H. Harriman, head of the Northern Pacific roads, had battled over ownership of various railroads. At times federal courts had to make the decisions. Hill concentrated on expanding his rail lines in Washington state, a number of which were in Spokane.[3] Harriman worked at creating a monopoly of rail transportation in Oregon by adding branches to his Union Pacific and Southern Pacific lines, making western Oregon well covered with rail transportation. However, residents in central Oregon complained to Harriman because they had none, so he began to survey a route along the Deschutes River canyon and through the middle of the state.[4]

For a long time Hill had wanted to build a railroad through central Oregon that would eventually reach San

---

[2]Ibid., 57 and 60, and Hidy and Hidy, *John Frank Stevens, Great Northern Engineer,* 358.

[3][Spokane, Washington] *Spokesman Review,* Thursday, June 17, 1909, p. 1.

[4]*Railroad Age Gazette,* vol. 47: 905, November 12, 1909, p. 905.

Francisco. When John Stevens returned to work for the Great Northern, Hill's son, Louis W. Hill, was then president of that railroad. He decided it was time for his company to invest money in building a road through central Oregon. He believed a route could extend from the Columbia River, up the brown, treeless valley of the Deschutes River to the high plateaus in Oregon, as far as a little-known community named Bend. The river flows from Little Lava Lake about 40 miles southwest of Bend. It travels almost directly north to the Columbia River and joins it at Celilo, about 100 miles east of Portland. It pours down from an altitude of 4,000 feet, flowing northward across the great inland plateau where it is met by the Crooked River north of Bend, and continues twisting and turning 200 miles to join the Columbia River at Celilo. The Deschutes is too swift and dangerous to navigate. The Cascade Mountain Range runs along the entire west side of the territory, and the Blue Mountains are on most of its east side. However, the key piece of geography as far as Harriman and Hill were concerned was a 156-mile section of the Deschutes canyon north of Bend. In places it is wide enough to accommodate only one railroad line. The Northern Pacific Railway had a line running southwest from Spokane to Pasco, Washington, near the westward turn of the Columbia River. Hill had built a new line running along the north side of the Columbia from Pasco to Portland, with plans to cross the Columbia at Wishram, Washington, to Celilo Falls, Oregon. The line began operating in 1908, and was named Spokane, Portland & Seattle (SP&S).[5] Both companies had agreed to use those two lines for a common route into Portland. They would travel on the Northern Pacific from

---

[5] *Spokesman Review*, February 26, 1908, p. 2, and November 21, 1908, p. 16.

Spokane to Pasco, and then on Hill's SP&S into Portland. They built two large steel bridges across the Columbia from Vancouver, Washington, to Portland. The NP would use one, and the GN would travel on the other. A controversy had arisen over who was going to own the new railroad. A legal battle took place, and the court declared James Hill and the Great Northern as official owners. When Hill hired Stevens, he made him president of the SP&S, plus two subsidiary roads known as "Hill's Lines."

Stevens went right away to Portland. He made secret automobile trips around the territory trying to determine where in central Oregon a new railroad could be built. He finally agreed with Louis Hill that the Deschutes River canyon was the only route available. He learned on his tour that some years earlier other persons had taken steps to build a railway along the Deschutes River. They had filed corporation papers under the laws of Nevada for the Oregon Trunk Line Railroad and done some surveying, but no construction. He also discovered that the person who owned controlling interest in the defunct Oregon Trunk Line, William Nelson, lived in Portland.

John Stevens had some plans in mind, but he knew if those plans were revealed too soon, he would have a tremendous fight on his hands. He went to Portland incognito and pretended to be a genial sportsman named John F. Sampson. He appeared to be greatly taken with the area as a sportsman paradise and bought options on land and ranches in the Deschutes canyon area. Then he approached William Nelson and asked to buy his stock in the Nevada railroad charter. Although surprised at the request, Nelson made a deal with him. On September 6, 1909, Stevens, disguised as Sampson, met Nelson in a Portland city park on a rainy night. Stevens paid $150,000 by check to acquire Nelson's stock in the Oregon Trunk Line. He then registered the transfer in Washington, D.C., and obtained a new charter for the Oregon Trunk Line to build a railroad as far south as Klamath Falls, Oregon.[6]

He hired a crew of engineers and had them quietly set stakes along the old Trunk Line that had been partially surveyed. He employed one of his former Panama staff members, Ralph Budd, to be chief engineer of the project. He also obtained the services of Ralph Modjeski, who later designed and built some of America's greatest bridges, including the Delaware River Bridge between Philadelphia and Camden, New Jersey, in 1926, and the Mid-Hudson Bridge at Poughkeepsie, New York, in 1930. Before his death in 1940, Modjeski developed the preliminary plans for the San Francisco-Oakland Bay Bridge. The Porter Brothers were the contractors in charge of construction of the new railroad line. The first indication that the Harriman people had competition on the Deschutes was dirt flying from excavation of a railroad grade on the east side of the river, where they were also working. Stevens cast off his secrecy and announced through the press that the James J. Hill line was going to build a railroad from Celilo on the Columbia River south to Bend, a distance of 165 miles. They would build a bridge across the Columbia from Celilo to Wishram and connect with the SP&S railway.[7] Stevens' work crew began construction on July 23, 1909. Material and laborers began arriving at The Dalles, Oregon, on the Dallas, Portland & Astoria Navigation steamboat *Bailey Gatzert*. Four-

---

[6]Stewart H. Holbrook, *The Story of American Railroads* (New York: Crown Publishers, 1947), 184–85.

[7]Hidy and Hidy, "John Frank Stevens, Great Northern Engineer," 358–59.

DESCHUTES RIVER CANYON

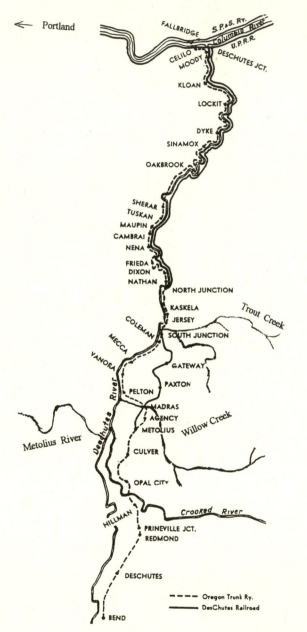

The Caxton Printers, Caldwell, Idaho.

horse teams with wagons then hauled them to the Porter camps.[8]

Meanwhile, Harriman's crews had also been preparing to construct their rail line in the Deschutes River canyon. They had a main railway, the Oregon Railroad & Navigation Company (OR&N), that paralleled the south side of the Columbia River from Portland to Umatilla, where it branched into Washington and central eastern Oregon. From the OR&N they had built a branch line that extended sixty-nine miles south from the Columbia to Shaniko. It roughly paralleled the Deschutes River varying from one to twenty miles east of it. Over this branch they sent box-cars loaded with horses and mules, flatcars of timber and rails, and coach-loads of laborers, and then hauled them forty miles upriver by wagon from Shaniko. George W. Boschke, who built the famous sea wall at Galveston, Texas, was the engineer in charge of the Harriman project. The Twohy Brothers were the contractors.

Groups of surveyors and workmen representing both Hill's and Harriman's interests were in the canyon by August 1909. Each desired to build on the side of the river that offered the most advantages. Their maps covered twelve miles of the same land. A clash was inevitable. Both were working on the east side when Harriman's crews discovered their competitors were Hill's men. That began a bitter battle! Immediately, Boschke, chief engineer of Harriman's line, known as the Des Chutes Railroad, challenged the validity of some earlier surveys made by the original promoters of the Oregon Trunk Line, which the Great Northern now owned. In addition, they sent a party of workers to physically force Stevens' contractors out of the

---

[8]Gaertner, *North Bank Road, The SP&S Railway*, 101.

canyon. A serious fight took place, but Harriman's employees failed to dislodge Hill's crew and withdrew from the fracas. The survey controversy showed that the Oregon Trunk Line was to build on the west side of the canyon.

Because some of the surveying on the Oregon Trunk Line's charter, which Stevens had bought, was being questioned, his attorneys, Carey and Kerr, moved the case from the state to the federal court, a privilege they had because the charter was of a Nevada state corporation, and they were working in another state. On August 24, the United States district court at Portland issued an injunction restraining representatives of the Des Chutes Railroad from going on that part of the right-of-way that had been surveyed by the original promoters of the Oregon Trunk Line. Hill's forces under Stevens had won. Emotional tensions were high among the settlers in the Northwest because the two railroads were allowed to build simultaneously in a narrow canyon. As a result, the Oregon legislature put a curious provision in the charter it granted to the Oregon Trunk Line. Each company official had to:

> —solemnly swear that I have not fought a duel nor sent or accepted a challenge to fight a duel nor been a second to either party—and that I will not be so engaged-in or about any such duel during my continuance in office.[9]

No duel was fought, but violence and strong-arm tactics were common. Despite the result of court action, work on both railroads continued. About 3,000 of Stevens' workers were encamped on the west side of the river, and around 3,800 of Boschke's men occupied the east side. It would have been a gigantic task to carve out just one railroad along a river hemmed in by towering walls, some ris-

---

[9]Frank Donovan Jr., "Canyon War," *Railroad Magazine* 67 no. 24 (October 1956): 24–25.

ing to a height of 2,000 feet, but with two railroads being
built simultaneously in that restricted area, the tension kept
mounting.

Stevens had purchased three farms along the canyon, one
of which was crossed by a wagon road constructed by Twohy
Brothers to supply their camp at Horseshoe Bend. One
time, Johnson Porter saw a long line of Harriman's wagons moving over the rough mountain road they had built.
He could see they would have to cross the land of his company's farm, so he hurriedly posted a sign which read: "No
Trespassing, Porter Brothers." One of the other farms contained a spring which supplied water for the Harriman crew.
Porter Brothers put up a sign there which stated: "No Water
To Spare." The contractors put sub-contractor Harry Carleton and 75 of his Italian laborers to guard the water.
Boschke filed a complaint which brought the sheriff, his
deputies, and a judge to view the situation. The judge gave
an order to give Harriman's wagons passage. When they
started through, Carleton and his men (probably ignoring
the judge's order) charged the intruders, which started a
fracas. After it was over, Carleton and several of his men
were arrested. A messenger from Porter Brothers arrived
shortly thereafter, ordering access be given.[10]

Stevens' Oregon Trunk Line followed the west bank of
the Deschutes to a point four miles north of Sherar where
the river makes a horseshoe bend. There the line crossed
to the east side in order to avoid a long detour. It arrived
close to where Harriman's men were working. At that point
it was necessary for the two railroads to excavate for tunnels. Both began digging and blasting within one hundred
yards of each other. The Oregon Trunk's dynamiting pelted

---

[10]Gaertner, *North Bank Road*, 102.

the Des Chutes Railroad men with so many rocks they had to temporarily leave their work site. Also, Boschke received a telegram, purportedly from Galveston, which stated: "The sea wall has broken." He realized at once it was a fake and threw it away, saying that he had built the sea wall to stand, and he knew it had. The OTL continued through their short tunnel and returned back to the west side, easing the tensions.[11]

E. H. Harriman died on September 9, 1909, and Judge Robert Scott Lovett assumed stewardship of the Union Pacific lines. Harriman's influence was estimated to extend over 60,000 miles of railroad tracks and resulted in better public railway service.

On March 2, 1910, the Oregon Trunk Line's bridge across the Columbia River at Celilo was authorized by an act of Congress. It was to consist of twenty-four spans with a total length of 3,348 feet. All piers for the many short spans would rest on exposed rock during the periods of low water. Toward the south end of the bridge a draw was to be built to handle traffic emerging from the Celilo Canal. Bids for the span were closed in April, and actual building began in July. Until construction was finished, traffic across the Columbia was carried by the 165-foot steamer, *Norma*, which was used with two barges fitted with rails for the ferry. One barge could carry eight rail cars and the other six. About 80 to 100 rail cars could be handled each day.[12]

Following legal controversies after the trouble at Sherar, an agreement was made between the two railroads to have the Des Chutes stay on the east side of the canyon and the Oregon Trunk on the west side as far as milepost 75 at North

[11]John Roger Twohy, *Ten Spikes To The Rail* (Jenner, Calif.: Goat Rock Publications, 1983), 67.   [12]Gaertner, *North Bank Road*, 105.

Ferryboat *Norma* used by James Hill to haul men and supplies
from the SP&S Railway at Wishram, Washington,
across the Columbia River to Celilo, Oregon.
Photo by Edwin D. Culp.

Junction. That area was in a very narrow section of the
canyon where there was scarcely room for a single track. A
homesteader, H. J. Smith, had proven his claim to the land
and had sold it to the Des Chutes line. At that point the
Oregon Trunk crossed over to the east bank and insisted
on retaining its earlier survey rights, which had included a
dozen miles on that side. On the west side was the Warm
Springs Indian Reservation. The road's management did
not want to get an easement through the reservation,
because there was no way to get to Bend from that side.

The only route was over Smith's property. Stevens—who held prior rights because he bought the charter of the original Oregon Trunk railroad, which allowed it to extend through that valley—quickly obtained an injunction to restrain his rival from entering the canyon. Crews from the Des Chutes ignored it and approached it from one end, and Stevens' crowd approached from the other. Both sides were armed and advanced toward each other like military troops moving into battle. (It could have been a bloody fray.) At this exigency, Robert S. Lovett, head of Harriman lines, and Jim Hill went into a personal conference in New York.

They reached a compromise, which gave the Oregon Trunk a right-of-way through the Smith property, permitted it to lay the tracks through the disputed canyon as far as South Junction, and allowed them to construct a bridge over the OR&N at Celilo. The Des Chutes was given rights to use the Oregon Trunk tracks from mile post 75 at Northern Junction to a point 500 feet south of Redmond. Rental was to be paid by them on a percentage of construction costs with taxes, maintenance, and operating expenses apportioned on a rail car mileage basis.[13] In January 1911, the Des Chutes Railroad requested from Hill permission to extend its use of the joint line from Redmond to Bend, which Hill granted.

Hill's crews completed laying tracks on the east bank to Pelton where the route left Deschutes canyon and turned into Willow Creek canyon at Madras. The Oregon Trunk Line arrived there in February 1911. From Madras it continued on to Metolius, a table-land level of 2,518 feet, and then on to Crooked River, three miles south of Opal City, arriving there in April. At Crooked River the crews encoun-

---

[13]Donovan, "Canyon War," 25.

tered a deep 350-feet chasm that had to be bridged in order to continue to Bend.

Effective May 11, Stevens resigned as chief executive officer of the Spokane, Portland & Seattle; Oregon Trunk Line; and other affiliated railroads. He turned completion of the project over to his chief engineer, Ralph Budd. Stevens had organized the campaign and fought the battle nearly to victory.[14]

Steel for bridging the Crooked River had not yet arrived by April, and when sufficient material finally came, work commenced in June. Ralph Modjeski had designed a graceful cantilever bridge on which workers laid the rails in September. It was an impressive structure, spanning a columnar basalt canyon 350 feet wide and 320 feet deep. The bridge consisted of a 340-foot, two-hinged arch span with a 60-foot deck-plate girder at each end, and was 320 feet above water level. At that time it was the second highest bridge in the U.S., after the Southern Pacific's Pecos River viaduct, which stood one foot higher. Work progressed more quickly on level terrain, and the Oregon Trunk reached Redmond on September 21; on September 30, its line ended at the final destination of Bend.[15]

On October 5, 1911, Jim Hill, accompanied by Louis Hill and Carl Gray, former vice president of the St. Louis & San Francisco Railroad who had replaced Stevens as president of the SP&S and the Oregon Trunk railway, took part in the ceremony of driving the final spike in Bend, Oregon. An iron spike was removed, and a golden one was produced. Seventy-three-year-old Hill drove it all the way into the railroad tie with two blows. The gold spike was later

---

[14]Gaertner, *North Bank Road*, 107; and Hidy and Hidy, "John Frank Stevens, Great Northern Engineer," 361.　　　[15]Ibid.

Oregon Trunk Railway Bridge over Crooked River.
Oregon Historical Society.

removed and taken to Portland to be displayed. In a speech to the people present, he said:

> We mean to get along with our neighbor, the Union Pacific. When they get in a tight place we are going to extend a helping hand

and if we get in a tight place we will call on them. We won't make faces across the fence.[16]

Meanwhile the Des Chutes Railroad crews had extended their line to reach Madras on June 6, 1911. They were six weeks behind the Oregon Trunk in spite of having twenty miles of track laid when the Oregon Trunk began theirs. In order to continue to Metolius they had to build a bridge to cross both Willow Creek and the Oregon Trunk line. The bridge, which they completed in August, was 1,050 feet long and 250 feet high. The line moved on to Metolius, arriving in September. Oregon Trunk crews were finishing their bridge across the Crooked River chasm when the Des Chutes crews arrived. The latter, having neither the desire nor the finances to build their bridge across the deep gap, decided not to do it and ended their railroad construction there.

The Des Chutes Railway made an agreement with the Oregon Trunk Line to use the Oregon Trunk rails from the north end of Metolius to Bend, a distance of forty-one miles. Passenger trains from the north would be consolidated at Metolius and run into Bend. North-bound trains would be split at Metolius to continue on their respective railroads.

The first regular passenger train pulled into Bend on October 30, 1911. It traveled from Wishram, across the Columbia via the steamer *Norma*, and on the new Oregon Trunk railway. On January 5, 1912, the Columbia River bridge was completed, and trains started using it two days later.

With John Stevens' leadership, Jim Hill's promise to build a railroad through central Oregon was fulfilled. From Wishram to Bend, nearly nine million cubic yards of rock and hardpan were handled. It was necessary to blast nearly all of the rock in order to move it. With the exception of

---

[16]Donovan, "Canyon War," 25, and Gaertner, *North Bank Road,* 108.

some 200,000 yards of material near the north end that were moved by steam shovel, all excavation was accomplished by pick, shovel, and wheelbarrow. It was one of the largest pieces of railway construction done in this manner in recent years. The Oregon Trunk Line built ten steel bridges and nine tunnels. The Des Chutes Railroad constructed two bridges and five tunnels. Considering both lines, the tunnels, which varied in length from 415 to 800 feet, extended a total of 3,968 feet. They were lined with timber and later with concrete.[17]

When railway construction was first contemplated through the Deschutes River canyon, papers were filed with the federal government for dam sites for power development. The dams were figured to be 140, 100, and 60 feet high and located about 2, 20, and 40 miles, respectively, south of the confluence of the Deschutes and Columbia Rivers. The first and last sites were filed by private corporations, while the middle site was reserved by the U.S. government. It was necessary for both roads to locate above the proposed high-water levels to avoid possible future trouble.[18] Eventually, two dams were constructed on the Deschutes River, the Pelton and Round Butte.

In 1931, fifteen years after Jim Hill's death, the Oregon Trunk Line was constructed from Bend to Bieber, California. There it connected with the Western Pacific railroad to continue into San Francisco. That completed Hill's dream to build a line through central Oregon that would connect with the ports of California. What had been a sort of backwoods fight in an obscure canyon had developed into the inception of a new and important route to the Golden Gate.

[17] *Railway Age Gazette* 47, no. 20 (November 12, 1909): 906, and vol. 52, no. 12 (March 22, 1912): 680–683.

[18] Ibid., March 18, 1912.

6

RAILROADING IN RUSSIA
DURING WORLD WAR I

After returning to New York, Stevens worked for five years, 1912–17, in private business. Some of his projects as a railroad consultant included making a complete survey of the Spanish railroads for the National City Bank, Kuhn-Loeb Company, and J.P. Morgan, and serving as an advisor for the New York subway system.

When he knew his old boss and friend, James J. Hill, was coming to the city, he would go and visit him. As Stevens was leaving after one of his visits, Hill followed him to the elevator, put his hand on Stevens' shoulder, and said, "John, is there anything in the world I can do for you?" "No, Mr. Hill," Stevens replied, "you have done a lot for me, and I want you to know how deeply I appreciate it." Hill continued, "Well, if there ever is, you know where to find me."[1]

James Hill, the great railroad genius, died a few months later, on May 29, 1916. Stevens felt his death as a personal loss.

During the five-year span of Stevens' work in New York, events occurred in Europe that would increase his railroad service. In 1908 Austria-Hungary had annexed Bosnia-Herzegovina; because they feared Serbia would take over

---

[1]"An Engineers Recollection," 37.

that territory, they attempted to control Serbia in this way. Those policies led to the assassination of Archduke Francis Ferdinand, heir to the Hapsburg throne of Austria, and his wife, the duchess of Hohenberg, on June 28, 1914. They were visiting Sarajevo, Bosnia, a provincial capital in the Austro-Hungarian empire; while riding in a car, the two were shot by a young Bosnian nationalist student. Considering this a Slavic challenge to the Teutonic people, Austria-Hungary called upon the Serbian government to publish a regretful and semi-apologetic declaration on the front page of its official journal. In addition, they issued demands with which Serbia was to comply, two of which Serbia refused. One was that Austro-Hungarian government officials were to assist Serbians in suppressing all anti-Austria material in their publicity and public instructions. The other was that Serbia was to take judicial proceedings against persons accessory to the assassination plot, with Austrian officials taking part in the investigations.

Serbia was willing to comply with the other demands; these two, they believed, were calculated to subvert their sovereign independence and were contrary to their constitution. They requested permission to submit them to the Hague Tribunal. However, Austria announced that nothing short of complete acceptance would suffice. Having earlier received a promise of military support from Germany, they declared war on Serbia on July 28, 1914. Russia sided with Serbia and threatened to intervene against the two Teutonic countries. France joined with Russia and Serbia. Britain, being a protector of Belgium, remained neutral.

On July 29, 1914, Germany declared war on Russia, Serbia, and France. In order to invade France, Germany violated international law by sending troops across neutral

Belgium. This brought Britain to the combat, and that country declared war on Germany. The United States stayed neutral. Germany began to use extensive submarine warfare, and in May 1915, it sank the British ocean liner *Lusitania*. About 1,200 men, women, and children were drowned, among them some 146 Americans. German subs also sank the American ships *Gulflight* and *Nebraskan*. President Woodrow Wilson sent a complaint to Germany denouncing the sinking of the ships as a violation of international law. He declared that the United States would not "omit any word or act to maintain the right of neutrals to travel on their legitimate business anywhere on the high seas."[2] However, Germany continued its submarine destruction. In January 1917, they boldly proclaimed a policy of unrestricted submarine warfare in the waters around the British and French coasts: All ships found in this zone, neutral and belligerent, armed or unarmed, would be sunk.[3]

On April 6, 1917, the United States declared war against Germany and later against Austria-Hungary. By then, Germany had invaded Russia and made steady advances. The Russians not only had to fight well-equipped and well-trained enemy soldiers on its eastern section, but they were hindered also by their decrepit railroad system, which consisted of far-flung trunklines and a tremendous amount of worn-out equipment.

The German navy had blocked entrance to Russia's Baltic Sea port south of Finland in the Finland Gulf, and the Black Sea port north of Turkey. This closed their two main shipping routes, and left two available: Archangel in the far north on the White Sea, closed by ice from October to May, and

---

[2]J. Salwyn Schapiro, *Modern and Contemporary European History (1815–1934)* (Boston: Houghton Mifflin Co., 1934), 731.   [3]Ibid., 733.

the far eastern port of Vladivostok on the Sea of Japan. At that seaport huge stacks of stranded imported freight crammed the docks, filled all available warehouses, and were piled out in the weather waiting to be distributed in Russia.[4] The deteriorating system of Russia's railroads caused anxiety among America's allies. When the United States entered the war, the Allied nations urged President Wilson to take charge of improving Russia's rail situation.

In addition to the Allies' pressure for the United States to aid Russian railroads was further information from an American correspondent of the *Chicago Daily News* and *London Times*. His name was Stanley Washburn, and he had covered Russian combat on the eastern front, reporting on a massive Russian retreat from Poland, and also a costly Russian offensive the following year. Washburn had gained exceptional knowledge of Russia's supply problems. He kept in close contact with his friend and employer, Lord Northcliffe, the owner of the *Times* who, along with other British leaders, was urging the Russian Provincial Government to request American management of the Trans-Siberian Railway as a means of improving supply and the military outlook. At one point Washburn had to return home to recover his health in an East Coast sanitarium but, desiring to use his influence to help the situation in Russia, he went to Washington, D.C. In the capital Washburn contacted Daniel Willard, president of the Baltimore & Ohio Railroad and chairman of the advisory commission to the Council of National Defense (CND). Willard listened to Washburn's comments and then requested Secretary of War Newton D. Baker, chairman of the Council of National

---

[4]Theodore Catton, "League of Honor, Woodrow Wilson, and the Stevens' Mission to Russia," University of Montana Master's Thesis, 1986, unpublished, 16.

Defense, to have Washburn brief the CND. Baker complied, and the correspondent related Russia's plight, indicating that it could seriously jeopardize the Allies' war plans. He also detailed plans for sending a railroad commission to inspect the entire railroad line from Vladivostok to Petrograd (St. Petersburg), the capital, and assess the needs of the railroads for locomotives, cars, and operators, which the United States would supply. After the meeting Baker discussed the idea with President Wilson, who told him to consult with the secretary of state, Robert Lansing. The president also asked Baker to inquire through the American ambassador in Petrograd, David R. Francis, whether such a commission would be welcome with the new Russian government.[5]

On March 15, 1917, a month before the United States declared war on the Central Powers, a Russian revolution forced Czar Nicholas II to abdicate the throne. A provisional government led by George Lvov as prime minister and Alexander Kerensky as minister of justice was instituted. President Wilson was anxious to establish and maintain a good relationship with the new government.

Secretary Lansing was enthusiastic about a U.S. railroad commission, but Ambassador Francis cabled from Petrograd that Russia had refused the offer. In a later telegram he explained the reason for their refusal. The British ambassador to Russia had told two ministers of the government that the Trans-Siberian Railroad ought to be put under American operation. The Russians objected because they did not want the Americans to have virtual control of their railways, nor did the Provisional Government want American assistance to imply any special American privileges in Siberia after the war. However, following further

---

[5]Ibid., 20.

discussions, Russian leaders changed their minds. They agreed to accept a railroad commission and sent a formal, written agreement stipulating their terms, which mostly emphasized their receiving material assistance. There was no mention of American operation of the railways, but in the minds of the Allies the main problem was the way the Russians were doing the operating.[6]

When Russia's first refusal to have a railroad commission was received, President Wilson decided instead to have a diplomatic mission, which would be more advantageous in encouraging democracy. He appointed a commission representing labor, business, military, and the YMCA. He asked former Secretary of State Elihu Root to head the mission. It was while the president worked on organizing the Root Commission that the Russians changed their minds and asked for the Railroad Commission. That caused Wilson to revert back to the original plan of sending a commission of railroad engineers. Now there would be two commissions, one diplomatic and one technical. Root was angry over the change, but there was nothing he could do about it.

Daniel Willard was assigned to select the railroad engineers. He picked his old friend, John F. Stevens, to lead the group. Other members of the commission were Henry Miller, a very able operating engineer; William Darling, a specialist in locating and maintaining railway lines; George Gibbs, an expert in equipment, maintenance, and repair shops; and John E. Greiner, an eminent bridge engineer.

During March and April 1917, Stevens was in New York and took no part in the Russian railroad deliberations. He would never have guessed that he would be called upon for the third time in his long railroading career to act as a high

---

[6]Ibid., 25, citing Russian Embassy to Department of State, undated.

Harriet O'Brien Stevens,
1854–1917.

representative of his government. However, at the age of sixty-four, he was still vigorous and energetic. His temperament suited the rugged, masculine life of a civil engineer, but it would not serve him so well in Russia, where he would need patience and diplomacy. He was vacationing in Bellaire, Florida, when Willard first contacted him. A few months earlier, on Thursday, January 18, 1917, his beloved wife, Harriet had died in their home at 875 Park Avenue, New York City. They had been married forty years, and he was still experiencing his sorrow and loneliness for her.

When he agreed to accept the appointment to head up the Railroad Commission, he didn't realize the short two-to-three-month mission Willard promised him would grow into a diplomatic tour of almost six years in Europe and Asia. Neither could he have foreseen the role he would be called upon to play as a U.S. representative in the higher echelons of Far Eastern politics.[7]

Secretary of State Lansing approved Willard's selection of railroad engineers, and appointed them to serve on the commission. In May he accredited them to the Russian government as "officers pro tem of the U.S. Department of State." On May 8, President Wilson summoned them to the White House for a briefing on their purpose in Russia. He told them they were not a political nor diplomatic commission. Their duties would be confined to advising and assisting the Russians in their transportation problems and to advising on such railway matters as they might suggest. Specifically, he cautioned them they were not going to Russia to ask, "What can the United States do for Russia?" but to say, "We have been sent here to put ourselves at your disposal to do anything we can to assist in the working out of your transportation."[8] On the previous day President Wilson had told Secretary Lansing that the Railroad Commission was to be entirely different from the Root Commission, and Stevens was anxious to clarify that point. He pressed Wilson for his assurance that his commission alone would have the authority to discuss the railway situation with the Provisional Government. He wanted to be sure that the Root Commission did not interfere with the Railroad Commission's work. However, the president gave no direct reply.

Stevens left the meeting dissatisfied. He was not impressed by the president's lofty rhetoric. He wanted def-

---

[7]Raymond Estep, "John F. Stevens and the Far Eastern Railways, 1917–1923," *Explorers Journal* 48 (March 1970): 14.   [8]Ibid., p. 15.

inite instructions and prerogatives, and was anxious about the commission's reception in Vladivostok, where their tour was to begin. Later, on a layover in Tokyo he discussed Wilson's briefing session with the British ambassador. Stevens told him that the president had given him no instructions except to offer his services to the Russian government and people, and to render them every possible assistance in the war against the common enemy. Perhaps feeling that his mission would call for some skillful diplomacy, Stevens added, "My business will be to do the work in Russia myself, and to make the Russians think that they are doing it."[9]

At Secretary Baker's request, Washburn was commissioned a major in the United States Cavalry. Baker ordered him to accompany the Railroad Commission to Vladivostok, where he was to transfer to the Root mission.

The Railroad Commission left Washington, D.C., by train on May 9. The staff that accompanied them consisted of E. P. Shannon, secretary; Franklin Reading, disbursing officer; Stanley Washburn, military attaché; Eugene Stevens (son of John F.), assistant; C. A. Decker and Leslie Fellows, stenographers; and F. A. Golder and Eugene Prince, interpreters.[10] They traveled across the country to Vancouver, British Columbia, where they boarded the steamship *Empress of Asia*, and after a drizzly three-day voyage, arrived in Yokohama, Japan. Following a brief layover, they boarded the Russian steamship *Penza* and sailed two days across the Sea of Japan, arriving at the Port of Vladivostok on May 31. A crowd was on hand to welcome them. Among the

---

[9]Catton, "League of Honor, Woodrow Wilson, and the Stevens' Mission to Russia," 32, citing Jacqueline D. St. John, "John F. Stevens; American Assistance to Russian and Siberian Railroads," University of Oklahoma Master's Thesis, 1984, 109.

[10]Joe Michael Feist, ed., "Railways and Politics: The Russian Diary of George Gibbs, 1917," *State Historical Society of Wisconsin* (Spring 1979): 181–82.

Members of the American Advisory Railroad Commission
in Russia. *Princeton University Library.*

greeters were a local executive committee of the Council
of Workers and Soldiers, the military governor, the mayor,
a Russian military general who expressed the people's will-
ingness to accept American assistance, and a representative of the provisional government, A. N. Mitinskii from
Petrograd. Also present were the traffic manager of the government railways and the chief engineer of the Chinese
Eastern Railroad.[11]

---

[11]Catton, "League of Honor," 36–37.

After an exchange of formalities, Mitinskii took the Americans to their accommodations, which consisted of a special train of three sleeper cars, a diner, an observation car for Mitinskii, and an office car for Stevens. The commission members settled into their quarters and then were taken on a tour of the three-mile-long harbor. As they cruised, they were amazed to see mountains of freight-rail shipments. Bales of cotton were exposed to the weather. Thousands of drums filled with nitrate fertilizer, crates, barbed wire, and steel rails were stacked along the docks. They looked at sheds overflowing with explosives and artillery shells. Some munitions were loosely covered with tarps, and others lay in the rain. After the tour the commission talked about what they had seen and concluded the situation was not the fault of the local port administration. The main problems existed inland along the Trans-Siberian Railway.

Later in the day, American lieutenant E. Francis Riggs, a military attaché to the Russian embassy, spent some time with Stevens briefing him on the political situation. Riggs informed him that to interpret the Russian attitude was difficult. They might be dragging their feet in an attempt to forestall their American presence in Vladivostok, or they might be trying to focus on Vladivostok in order to limit the work of the Railroad Commission to just the supervision of that port's administration. Riggs felt that the provisional government did not want the commission's advice concerning the Trans-Siberian Railway or the rest of Russia's railroads. The authorities in Petrograd believed the congestion in Vladivostok resulted from poor organization of port officials and not from the breakdown of operations along the railways. Therefore, they were interested in stationing Stevens' commission in Vladivostok, where they hoped it would act only as a channel for receiving locomotive shipments and other railroad supplies form the United States.[12]

At the close of Lieutenant Riggs' synopsis, Stevens was doubtful about administrating and clearing the port. He could accomplish little at Vladivostok if the real problems lay inland along the Trans-Siberian rail line. Yet, he wanted to know where the provisional government desired his commission to concentrate its efforts.[13] In a move to verify that the railway terminals were not the source of trouble, the commission went the next day to the nearby switchyards at First River. Railway workers at Vladivostok loaded boxcars in the port's cramped yards and then delivered them to First River where the trains were made up. They were loading as many as three hundred boxcars per day for shipment out of Vladivostok, but only one hundred cars were leaving First River daily. The port and yards around Vladivostok were clogging the flow of supplies to western Russia. The limiting factor was the shortage of locomotives and cars that returned from the West. Also, they observed that the important job of making up the trains and determining their destinations was handled by the Soviets.

In 1905 a rising tide of political discontent had resulted in a general strike throughout Russia. It began with railroad men and telegraphers, who cut off all communication and discontinued work. The stoppage spread to shipyards, factories, mines, and shops. The movement was led by a new organization called the Soviets, or Council of Workers Delegates. It consisted of representatives from the trade unions.

By 1917 the power of the provisional government rested in the Soviets, through which the Bolsheviks eventually

---

[12]Ibid., 38–39.   [13]Ibid.

gained control. The Soviet Executive Committee decided
what kinds of material would be shipped and in what quantities. They decided how far they would honor the requests
of the provisional government and how much food and fertilizer they would send instead of munitions and barbed
wire. Their interest was in supplying the people and farm-
ers with material for growing food. They wanted peace and
not military warfare.

During their inspection tours and meetings with Mitinskii and local railway officials, Stevens' group found they were
followed by a silent observer from the Soviets. Evidently,
the man wanted to know everything the Americans proposed. The local workers didn't want an influx of American laborers taking away their jobs or lowering their wages.
One of the commission members, George Gibbs, wrote in
his diary, "We would be certain to antagonize someone and
to cause misunderstanding and bad feelings."[14]

One project the Russians planned was construction of a
Baldwin Locomotive assembly plant at the head of the harbor. Stevens' commission visited the proposed site and found
it to be very adequate. They informed the Russians that
Baldwin Locomotive would want to supply forty or fifty
men to supervise the construction. The Russians agreed,
and Stevens promised to cable Baldwin for exact specifications for the buildings.

On Sunday, June 3, the ship carrying the Root Commission arrived in the Vladivostok harbor. It was a holi-
day, and the Soviet's Executive Committee was late in
assembling to receive them. Root and his men waited on
board. Stevens, on the verge of beginning his long journey
west, decided not to wait until they came ashore, as there

---

[14]Feist, "Diary of George Gibbs," 183.

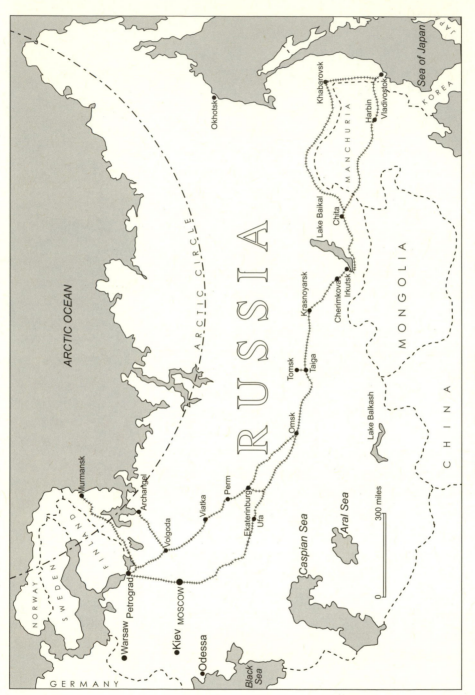

Chinese Eastern Railway in eastern Russia.
Based on map by John F. Stevens. *State Historical Society of Wisconsin.*

was some confusion between the two groups regarding their mission. In addition, Mitinskii was anxious to keep the Railroad Commission on schedule. He said it was important that they get to Petrograd ahead of the Root Commission. They hurriedly loaded into a special train and left Vladivostok on June 3 to begin a trip of 5,800 miles among the mountains, desert, and steppes that lay beyond. Washburn remained behind to join the Root Commission as Baker had ordered him to do.

The first part of their journey was a vital shortcut across northern Manchuria on the Russian-owned Chinese Eastern Railway. En route to Petrograd, the visitors planned to inspect railway facilities at most of the major terminals, and at Harbin in Manchuria, they had their first opportunity to do so.[15] They received a hearty welcome from the Russian military governor, General Dmitri Horvat, who was also general manager of the Chinese Eastern Railway. He was accompanied by Chinese government officials and a delegation of Mongolian princes. The commission examined the large shops where most of the locomotives and cars imported from the United States were assembled. They considered it most inefficient for parts of American locomotives to be unloaded from ships in Vladivostok, hauled by rail to Harbin to be assembled, and then driven back to Vladivostok to pick up their first loads of westbound freight.

In Harbin General Horvat treated the Americans to a sumptuous lunch at the Eastern Railway Club, and then accompanied them on a steamboat down the Sungari River for an examination of the long-girder railway bridge. Finally, as Stevens' commission prepared to depart, the train

---

[15]Estep, "John F. Stevens and the Far Eastern Railways, 1917–1923," 15.

carrying the Root Commission pulled in behind their own, so the railroad engineers had to attend another banquet and round of formalities. Numerous Russian generals were present and called on Stevens to give a speech. In his remarks he praised the operations of the Chinese Eastern line.[16] Finally, the Root Commission's special train departed to continue its journey. The train carrying Stevens' commission left four hours later.

The next day the railroad commissioners arrived at Manchuria Station, a small, bleak settlement on the Russian border that marked the juncture of the Chinese Eastern line with the Trans-Baikal railroad. Here they encountered major congestion—five hundred loaded boxcars standing abandoned on the sidings. Many of these were American-made cars, apparently loaded for the first time but left stranded for eight months. The problem lay with the Trans-Baikal, which was behind schedule in dealing with maintenance as well as the movement of equipment.[17]

Their train continued on to Sungari, an area surrounded by bridges that the commission stopped to inspect. They were impressed with the permanent steel bridges built on huge wedge-shaped abutments that pointed upstream to deflect the ice floes in the spring. There were also a number of wooden bridges, some barely capable of supporting the American locomotives. The engineers found the ballasting to be poor in places and the water and cooling stations crudely made.

At Chita, capital of the Chita province with a population of 80,000, mostly Chinese, the commission came upon the crudest facilities they had yet seen on their trip. The

---

[16]Catton, "League of Honor," 46–47.
[17]Ibid., 48.

repair shops were old and poorly designed. Their equipment consisted of only light tools, a wheel-and-axle lathe, and a small brass foundry. The roundhouse contained neither a crane nor a drop pit. Only light repairs could be performed. The engine house had all the broken-down locomotives it could accommodate, and more disabled ones stood outdoors in the weather. Inspecting the coal pit just beyond Chita, the railroad experts discovered the primitive method of loading coal into the engine-tender from small buckets lifted by a hand-operated beam in a see-saw manner. The engine was uncoupled form the train and had to run some distance to the coal yard. The procedure took about two hours.[18]

From Chita the Americans continued on the Trans-Baikal Railway to the city of Irkutsk, passing the Root Commission's train along the way. As their train pulled into the station, it was mobbed by a crowd of undisciplined soldiers, ex-convicts, and political exiles. They were prevented from boarding by the ten soldiers who served as armed guards. The Americans discovered that soldiers all over Russia were boarding trains at will. They crowded into compartments, slept in aisles, or sat on the roofs. Since the overthrow of the czar in March of that year, the Russian people believed themselves to be free, and freedom meant license to call for complimentary services, so the railways had stopped charging fares for armed soldiers.

While the Railroad Commission's train was refueling, the Root Commissioners arrived. A crowd gathered around their train, and two members of the Root Commission, James Duncan, the labor representative, and Charles Russell, a socialist, made speeches proclaiming America's

---

[18]Ibid., 51.

enthusiasm for the democratic principles of the Russian Revolution. The crowd received their talks with repeated applause. Meanwhile, Stevens went to visit Root, who was riding on the ex-czar's imperial train. He showed Stevens the desk at which the czar had signed his abdication.[19]

On June 13, 1917, ten days after leaving Vladivostok, the train carrying the Railroad Commission finally arrived in Petrograd. Train platforms were choked with refugees trying to get out of the city. Families with all their possessions had been waiting for days to get a train. Petrograd was desperate for transportation. It also needed coal for the factories and food for the hungry population. Twenty percent of the factories had closed down for lack of fuel, and bread lines had grown a block long. Soldiers strutted among the people carrying rifles and ammunition belts. They heeded no one. U.S. ambassador David Francis and his staff were at the station to greet the Stevens commission. They were accompanied by Russian minister of transport Nikolai Nekrasov, and a delegation of his railroad officials. Stevens read to them a prepared statement announcing the commission's intent to assist the Russian government in railway matters by placing at its disposal America's technical skill and industrial resources.[20]

After a lukewarm reception, the commission members were taken to their accommodations, which turned out to be a shabby hotel near the train depot. It was ironically called the Select. Francis apologized for its condition and explained the Russians had planned to put them in the Winter Palace, but it was now reserved for the Root Commission. The other lodging, Hôtel de l'Europe, was overcrowded.

The next day the Railroad Commission went to the

---

[19]Ibid., 53–54.   [20]Ibid., 60–61.

American Embassy for a conference with Ambassador Francis. The Root Commission arrived in Petrograd the same day. On the following morning railroad commissioners returned to the embassy to meet Nikolai Nekrasov. He had been a civil engineer and professor in the Tomsk University before his political appointment. In the discussion with him, the Americans realized their efforts at reorganizing Russian railroads would be extremely difficult.[21]

That evening Francis took all five commission members for a drive in his Ford touring car along the Neva River. For Stevens it was a chance to observe two Petrograds. One was the graceful city of canals, bridges, Italian-style architecture, and soft, pastel shades of color. The other was a city in revolution, festooned with red government banners but overcrowded with refugees, army units, and peasants suffering from severe shortages of food and fuel.

On June 15, three days after he arrived in Petrograd, Stevens fell ill with erysipelas, a skin infection caused by streptococcus that produces a painful rash with fever and chills. He was hospitalized for nearly a month, and Henry Miller served as head of the commission during his absence.[22]

Under the Czarist government the Ministry of Transport had centralized control over the railroads. After the March Revolution, that government had collapsed, along with control of transportation. The new provisional government was unable to restore that authority because during the rebellion, railroad employees had formed labor unions that usurped the power of government officials. The Americans' handicap was that they could work only through

---

[21]Feist, "Diary of George Gibbs," 187.
[22]Estep, "John F. Stevens and the Far Eastern Railways," 15; Catton, "League of Honor," 80–81.

the Ministry of Transportation, which was influenced by the labor unions.[23]

Daily sessions were held between the American Railroad Commission and Russian transport officials, including locomotive engineers of the various railroads. The meetings were laborious and discouraging for the Americans. Transport Minister Nekrasov was busy attempting to form a political coalition and often would cut short discussions in order to rush away and attend cabinet meetings. The commission soon discovered him to be long on making promises and short on fulfilling them, and most matters drew much talk rather than being acted upon. It was also difficult to carry on conversations through interpreters.[24]

After a week of discussions, the commission decided to present specific recommendations to the Russians. Each member organized data for improvements and discussed items related to his specialty. George Gibbs emphasized the necessity of maintaining locomotives and freight cars. He stated that production of passenger cars would be curtailed and assembly lines converted into repair shops. He recommended that the United States supply two thousand locomotives. Henry Miller urged the Russians to adopt the American railroad system of operation; each railroad should divide into divisions of 300 to 400 miles long; each division should have its own superintendent, dispatchers, trainmaster, and telegraph and telephone operators; each train engine should cover 150 miles per trip instead of the 75 miles then scheduled.

The commission recommended also that the capacity of the Trans-Siberian Railroad be increased by upgrading the coaling and watering stations. Parts and replacements

[23]Catton, "League of Honor," 72–75.   [24]Ibid., 79–80.

could be supplied by the United States to keep them functioning properly.

From his hospital bed, Stevens cabled Daniel Willard, chairman of the advisory commission to the Council of National Defense, requesting that he donate used locomotives and cranes left from building the Panama Canal. He thought they should be donated to the Russians to eliminate any suspicions that the Americans were trying to push off second-rate material at high prices. He hoped his large impending equipment order would get the Ministry of Transport to make the organizational changes on the railway.[25]

Eli Root's diplomatic commission in Petrograd was also working to carry out its mission. Root asked Stevens about speeding up Russian orders for locomotives, which had been requisitioned in May, by shipping some already in stock. The inquiry perturbed Stevens because he thought if the Root Commission got into the railway situation, it would undercut his authority and hamper his mission. His reply was that it was impracticable because it was too difficult to convert them to the Russian track-gauge. He said that the Wilson administration would not fill any locomotive orders until his commission made its recommendations. He did not want Root interfering.[26]

Russian officials received the Railroad Commission's recommendations for improving the operation of the Trans-Siberian line. They responded with a disappointing report. On matters of material improvements they consented from the standpoint of "technical realization only." They knew the purchase of new equipment was needed, but the financial question was beyond their scope. However, they said

---

[25]Ibid., 81–84.   [26]Ibid., 84.

they would submit the commission's proposals to Nekrosov for his consideration. He didn't respond to the commission. The commission left pending the crucial matter of funding and purchasing locomotives and railroad cars. This prevented Stevens from placing orders for the equipment.[27]

Deeply discouraged by the Russian report, Stevens blamed the failure on meddling by members of the Root Commission. He wrote an angry letter to Eli Root, accusing him of interfering in transportation matters and undermining the Railroad Commission's abilities to make improvements on the Trans-Siberian. He stated, "Any attempted interference by any other commission . . . would be a serious handicap" to the Railroad Commission.[28]

Stevens left the hospital to confer with members of his commission. They decided that Russian officials needed to be stirred up and some commitment needed to be obtained from Washington. Stevens cabled a message to Daniel Willard: "This whole nation [is] imbued with the one idea that additional engines and cars are necessary. [It is] expedient for moral effect [that] action be taken."[29] The next day he cabled Washington urging the government to order construction of 40,000 freight cars and 2,000 locomotives for Russian use. He also asked that the U.S. send an additional 500 locomotives immediately so that the stockpiles in Vladivostok could be moved east. In another cable to Willard, Stevens stated: "A prompt confirmation of this program by our government is valid to the Allied cause and the aims of the Commission. Answer."[30] He announced those recommendations in a speech to the Russian railway officials July 4. He had not yet shaken his illness completely and had to return to the hospital that evening.

---

[27]Ibid., 85.

[28]Ibid., 85–86.

[29]Ibid., 86.

[30]Ibid., 87.

While the Railroad Commission members struggled to formalize their recommendations, the Root Commission was deeply concerned about the depth of the Russian people's disillusionment with the war. German propaganda portrayed the Russian soldier as cannon fodder for the British, and this had a demoralizing effect. Also, commission members were alarmed by the ineffectiveness of their own efforts to publicize the United States' friendship and help. They wanted to increase American propaganda as a means to improve the situation. They intended to do this by conducting a motion picture campaign that would cost $5 million. However, the funds were not forthcoming. Disappointed, Eli Root and his commission left Petrograd on July 8 to return to Washington, where he could take his urgent case for propaganda directly to President Wilson.

Transport Minister Nekrasov wanted the Railroad Commission to inspect the railroads in the Ukraine. Members began their assignment on July 5. Stevens was back in the hospital and could not go with them. The first day they traveled four hundred miles over the Nikolaevsk Railway southeast to Moscow. After two days there they continued over the Kursk Railway to Kharkov. Their route went northwest from there, through the Donets coal and iron mines region on the Ekaterin line to Kiev. They passed several Red Cross trains carrying sick and wounded Russians back from the front. Their July offensive attack had been stopped after an advance of just twenty miles. Proceeding westward from Kiev, the commission took a new north–south rail line along the war front into the small city of Mogilev.[31]

Having sufficiently recovered from his illness, Stevens was discharged from the Petrograd Hospital and joined the com-

---

[31]Feist, "Diary of George Gibbs," 190–91.

mission in Mogilev, arriving the evening before his fellow workers. A delegation of army officers met them at the station and took them to the headquarters of General Alexis Brusilov, Russian army commander-in-chief, who occupied a castle at the center of the city. Although he was extremely busy conducting military campaigns, Brusilov received them courteously. He asked the commission to discuss with his staff ways to improve transportation in the military zone.[32]

Russian officers met with the American railroaders in the railway car of the adjutant general. They handed the commission members maps showing railway lines beyond the enemy front that they expected to recapture, and also lists of equipment and railroad supplies that the army needed. They urged the commissioners to use their influence to get the provisional government to give prompt attention to items they had requisitioned. Those included everything from medical supplies to cannons. George Gibbs thought their requests were "staggering" and compared the officers "to a child waiting to be fed."[33]

Following their discussion with Brusilov's officers, the commissioners left Mogilev on a special train to return to Petrograd, arriving there July 12. On July 14, Stevens received word from Willard that 500 locomotives and 10,000 cars had been ordered from the factories. That constituted only a quarter of the amount Sevens had recommended. The other 1,500 locomotives and 30,000 cars were under consideration. Stevens and his colleagues went to the railway ministry to resume their daily meeting. They were dismayed to find that not a single thing that they and the Russians had agreed upon had been attempted, nor had any order been issue to do so. The minister of communications

---

[32]Catton, "League of Honor," 93.   [33]Feist, "Diary of George Gibbs," 192.

asserted that he had no power to act. In addition, two Russian engineers who had been appointed to oversee changes on the Trans-Siberian Railway suddenly resigned, and a third one attempted suicide because of the political crisis.[34]

*   *   *

As has been noted, following abdication by the czar in March 1917, the provisional government was established with Prince George Lvov as prime minister. The cabinet was composed of representatives of various political parties including:

* Constitutional Democrats, called Cadets, who championed ideas of liberalism and represented working people who had become converts to socialism;
* Mensheviks (Russian "Minority"), who believed the way to socialism lay only through democracy and, therefore, the way from capitalism to socialism would be through orderly, constitutional methods, not by violent uprisings;
* Social Revolutionists (S-R), who believed the peasant, not the workingman, was to have the chief role in the revolution. No revolution could succeed in Russia unless the peasant participated. Ownership of the land would be vested in a village community (mir), not in the state or in the peasant; and
* Bolsheviks (Russian "Majority"), who were the radical left wing of the government and favored an armed insurrection of the working class which would be followed by the dictatorship of the proletariat or military rule of the working class. For them no truly revolutionary party would ever gain power through constitutional methods. Influential members were Vladimir Ilyich Ulyanov, who assumed the name Nikolai Lenin, and Lev Davydovich Bronstein, who used the alias Leon Trotsky.[35] The Bolsheviks later organized the Communist party.

---

[34]Stevens Papers, Hoover Archives, University of Stanford, CA; St. John, "John F. Stevens; American Assistance to Russian and Siberian Railroads," 1–12.

[35]Schapiro, *Modern and Contemporary European History*, 762–63.

Several of the political parties began to start another revolutionary movement to achieve a social transformation rather than a political one for the nation. They advocated their long-standing slogan: "Peace, Land, Bread . . . All power to the Soviets."[36] In Petrograd new political soviets of soldiers, workingmen, and peasants began to increase and soon fell under control of the Bolsheviks. At the front lines military discipline was seriously undermined. Soldiers fraternized with the Germans and attacked their own officers, denouncing them as czarists and counter-revolutionaries. The army was disintegrating into rebellious mobs. Added to the upheaval was discouraging news that General Brusilov's military offensive had failed to kindle patriotism in the ranks of the troops. Bread lines had extended in the capital, and restaurants were closed. The Railway Commission was feeling the food shortage, too.

On July 16, 1917, all of the Cadet ministers resigned from the provisional government. Kerensky, a member of the Social Revolutionary Party, remained its prominent member and won popularity as a radical orator. In early August Lvov left the post of prime minister, and Kerensky succeeded him. He had started out as minister of justice, and was made minister of war and navy in May. After ousting Lvov in August, he assigned himself the post of premier and of minister of war and navy. The government cabinet then consisted of Mensheviks and Social Revolutionists. The Bolsheviks immediately opposed Kerensky's regime, calling it a tool of the imperialist Allies. They increased their efforts to get public demand for an end to the war. The country—defeated, starved, exhausted, and embittered—gave a willing ear to the Bolsheviks' agitation for peace with Germany.

---

[36]Ibid.

The day after the political uprising on July 16, rioting began. The American commissioners stayed in their hotel rooms as armored cars and trucks with machine guns drove back and forth on the streets. That evening fierce fighting occurred between the Cossacks and the rioters. Kerensky brought in loyal troops from outside the city who ended the rebellion.[37] A government-inspired rumor that Lenin was a German agent led to an attempt to arrest him, but he escaped and fled to Finland.[38]

The Railroad Commission's work was suspended through the rebellion. When it was over, they learned that A. Liverovskii had replaced Nekrosov as minister of transport in Kerensky's new cabinet. Hoping the new minister would be more cooperative than his predecessor, the commission submitted a full list of proposals for the railroads in the Ukraine and Siberia. The recommendations focused on the two crucial matters of redirecting coal traffic in Siberia and lengthening engine runs. There were also a number of minor suggestions. Liverovskii requested a few days to look them over. When the commission met with him nearly a week later, it became apparent that the new minister was using the same dilatory tactics as the former one. The Americans realized they should have submitted only the two items that mattered most instead of the whole report. For three days they were engaged in wearisome debates over water stations, clinker pits, and coal and wood chutes. Hoping to try another tactic in his approach to the new minister, Stevens endured those fruitless meetings.

On the evening of July 24, the American engineers were invited to a dinner at the Russian Ministry. Stevens gave

---

[37]Feist, "Diary of George Gibbs," 193; Catton, "League of Honor," 95–96.
[38]Feist, "Diary of George Gibbs," 194.

a speech in which he praised the abilities of the Russian engineers. Following the talk, the Russians and Stevens took turns lavishing praise on each other. Commission member W. L. Darling was amused by the spectacle, but George Gibbs felt discouraged and began to think their mission was useless.[39] He became ill, and after three days spent in bed, he had several talks with Stevens about the necessity of having a member of the commission return to America to push for definite action from the State Department. Stevens agreed and suggested that Gibbs go.[40]

Aftershocks of the mid-July political upheaval kept occurring in the government's cabinet. The Ministry of Transport's leader, Liverovskii, was replaced by K. N. Vanifantiev. George Gibbs became displeased with Stevens' domineering attitude and with the Russian Railway Department's stalling tactics, and found an ally in Ambassador Francis. Francis was jealous of Stevens and once complained to Gibbs that several times Stevens—who did not have a lot of confidence in Francis—had sent hot-tempered messages to Washington in code over Francis' signature.[41] Throughout those hostilities Stevens wouldn't admit defeat and worked on a plan to have the United States send railroad specialists to help with the work on the Trans-Siberian Railway. On July 27, Vanifantiev sent him a written consent approving his plan to supply the specialists. Stevens immediately cabled Daniel Willard and recommended the organization of a military unit of 129 railroadmen to make up two traffic divisions. Their responsibilities would be to educate the Russians in American operation of districts

[39]St. John, "John F. Stevens, American Assistance," 164.
[40]Feist, "Diary of George Gibbs," 195.
[41]Catton, "League of Honor," 99; Feist, "Diary of George Gibbs," 195.

under supervision of superintendents for handling railroad traffic. He made another recommendation to the Russian Railway Department to make use of an instruction cadre of the U.S. railwaymen trained in the operation of a service corps. From all the many discussions at the Russian Railway Ministry, those two recommendations were the most important.[42]

The next priority on Stevens' agenda was gaining approval for longer engine runs, an action that was necessary for the success of adopting better methods in railroad operations. Again his recommendation met opposition from the Russian board, and Stevens realized that orders for reorganization of the railways had to come from higher up than the Ministry of Transport. He asked Ambassador Francis to intercede for the commission, and the ambassador arranged for a meeting with Prime Minister Kerensky. Francis told Gibbs that he didn't want Stevens to speak for the commission on this occasion. At the meeting Francis stated the purpose of their visit and reviewed the recommendations on which the commission wanted action. Stevens then seized the initiative and told Kerensky that the American Railway Commission had been invited by the Russian government, and if the commission's recommendations continued to be ignored, such a situation "might be regarded as an insult to Russia's ally."[43]

Later that day, the Americans were informed that Kerensky had decided to place the Tomsk Railway under military control, appoint L. A. Ustrugov, deputy of the Ministry of Transport, to be commissar of the Trans-Siberian Railway, and adopt all the commission's recommendations.

---

[42]Estep, "John F. Stevens and the Far Eastern Railways," 16.

[43]Catton, "League of Honor," 104–5; Feist, "Diary of George Gibbs," 196.

Stevens and Francis were greatly encouraged that after two months of agonizing negotiations the commission had finally achieved a breakthrough.[44]

Stevens made arrangements for Gibbs and Greiner to leave Russia. He, Darling, and Miller planned to stay and see that the American methods and personnel were introduced smoothly onto the Trans-Siberian Railway. The two commission members left in a special car on August 14, 1917. The next day Stevens received a startling reprimand from President Wilson in a cablegram from the State Department:

> The President appreciates very highly what Mr. Stevens and his associates are doing in Russia but thinks it wise to remind Mr. Stevens that it is important that the impression should not be created that he or his associates represent or speak for the Government of the United States. . . . The President does not wish in this way to discredit assurances already given but merely to convey a very friendly caution for the future.[45]

Back in the United States, changes in situations were creating problems that affected the railroad struggle in Russia. Contrary to Stevens' efforts in Petrograd, no individual in Washington was charged with overseeing the program of railroad assistance for the provisional government. That project was meeting with large problems of production and supply. Ships had to be found to transport the materials from the West Coast to Vladivostok. War production in general was oriented toward trade with France and England, so American shipping needed to be reapportioned from the Atlantic Ocean to the Pacific. Railroad assistance to Russia had to be forced on a business community that was skeptical of the Russian market.[46]

---

[44]Catton, "League of Honor," 105.

[45]Ibid., 106, citing correspondence Lansing to Francis, 861.77/150a, August 15, 1917, Record Group 59, National Archives.   [46]Ibid., 124.

President Wilson was pressured by two members of his administration as to what action to take toward Russia's railroad situation. Secretary of Treasury William G. McAdoo was sympathetic toward Russia's needs. He realized the American economy demanded some coordination, so he proposed to organize the nation's resources, both financial and industrial, under the auspices of one agency attached to the Treasury Department and under his direction. He was ready to make a large loan to Russia so they could purchase their equipment. However, the proposal was dropped when the two countries failed to negotiate the size of the loan. Another plan for economic adjustment was put forth by the secretary of war, Newton D. Baker. As chairman of the Council of National Defense, he had more control over war production than any other cabinet member. He saw McAdoo as trying to grab power from the council and argued that the secretary's plan gave too much control to one man. The American people, he said, would not accept an "economic tsar."[47]

The policy that he recommended to Wilson was to focus the nation's resources primarily on American military buildup in France. In April 1917, he had been supportive of the plan to send Stevens' Railroad Commission to Russia. Because shortages of railroad equipment both in the United States and France had occurred by the summer of 1917, he had become skeptical of the effectiveness of railroad exports to Russia. It was obvious that Stevens' request for 2,000 locomotives and 40,000 cars in Russia would stretch American resources dangerously thin. President Wilson decided to follow Baker's plan, which ensured that France, not Russia, would receive the bulk of American locomotives and rolling stock. The president repeated that

---

[47]Ibid., 130.

the original purpose of the Railroad Commission was to discover by what means the United States might assist Russia, but from the beginning he had regarded railroad assistance as a minimal commitment. For him, railroad assistance was a technical matter resting on the Russian war effort. With the collapse of Russia's military offensive in July, it was becoming evident to Wilson that the provisional government lacked the people's support to prosecute the war. He believed the wisest policy was to limit the locomotive shipments to what the United States could afford to lose if Russia quit the war, and restrict American operation of the Russian railways to what the provisional government definitely wanted.[48]

Britain also had become disturbed over Russia's transportation system. In a letter to Wilson the British ambassador in Petrograd stated that Russia's political and military upheaval was due to the deplorable state into which the means of transportation had fallen. He proposed that the United States take over the job of repairing all rolling stock by staffing the repair shops with reliable labor. If the United States didn't do it, then the ambassador proposed that the only other solution would be to divide the Russian railways into sections with one of the Allies in charge of each section. The British proposal led Wilson to clarify his policy of assistance toward the Russian situation. By August his concern was to stabilize the nation's beleaguered forces of democracy and to use the American Railroad Commission to forestall an Allied intervention on the Russian railways.[49]

Delegates from Britain, France, and Italy met in Paris to discuss transportation problems, including the growing cri-

[48]Ibid., 134–38.   [49]Ibid., 145–47.

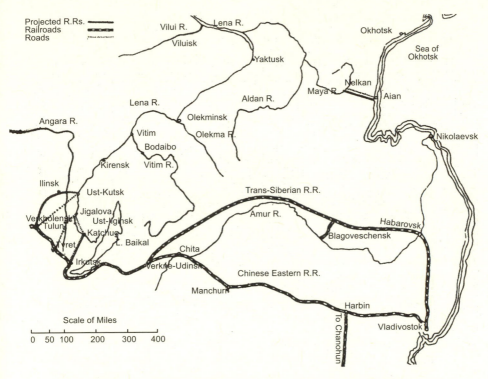

Lena region of Eastern Siberia showing location of railways.
Based on map by John F. Stevens in Trans-Pacific II, *February 1920.*

sis on the Russian railways. No representative from the Russian provisional government was invited to the conference,
and the Americans declined to attend. The Allies were concerned that with the Russian army breaking up and their
railroads on the point of total collapse, there could be a deep
German penetration into Russia. The northern Baltic city
of Riga seemed the likely place for invasion. If it fell, the
German army would march into Petrograd. Without rail
traffic the Provisional Government would be unable to

organize a defense of the capital. Another Allied concern was to keep the Murman Railway operational through the winter so Russia would not be cut off from Britain and the Atlantic supply line when the harbor at Archangel froze. That line ran from Petrozavodsk north to Murmansk. The Allies' conference closed with a pledge to the Russian government to provide all possible aid, both in material and personnel, for the rehabilitation of the railways, and with an appeal to the United States for more American involvement.

In Petrograd General Poole of the British military equipment section urged John Stevens to send one of his commission members to inspect the Murman line to make sure it was functional. Stevens consented, and the provisional government gave their permission for the inspection. W. L. Darling volunteered to make the trip. On August 7, he left on a special train with representatives of the British, French, and Russian staffs and several engineers. Along the way going and coming Darling noticed the railroad workers labored at their own speed and only when they wanted. The Russian officials paid them little attention and consented to whatever the workers requested. At Murmansk the general manager of the line assured the inspection group that the railway would be fully repaired by November 1. Darling became aware that if the railroad was going to carry traffic through the coming winter, the workers would have to concentrate on ballasting—filling the railroad beds with coarse gravel or crushed rock—building construction, and improving water supply. When he returned, he reported this to the commission. However, Russia did not ask the commission to offer recommendations regarding the Murman line. Both the Ministry of Transport and the railway department ignored the risky postponement of ballasting before winter arrived. Stevens did not press the issue because he preferred to concentrate the commission's efforts on the Trans-Siberian line.

Since Prime Minister Kerensky had previously approved the commission recommendations for improving the railroads, Stevens was anxious to take a second trip over the Trans-Siberian line with Transport Minister Ustrugov and his staff. With them he wanted to see how the new operating methods that his commission had recommended—longer engine runs and rotation of shifts in the repair shops—were to be adopted. Ustrugov delayed their departure by bringing up various excuses. Finally he told the commission to leave without him, and that he would join them in a few days. On August 24, Stevens, Miller, Darling, and the remaining commission's staff left Petrograd to begin another intensive survey of the facilities on the Trans-Siberian line. Ustrugov joined them in Siberia, and the group proceeded toward Vladivostok. On the way they met with railway officials and implemented changes in the new operations. Their journey lasted nearly a month, and they arrived in Vladivostok on September 22. Stevens was bedfast with an acute sinus infection and, according to Darling, was "too bull-headed to go to Japan to receive proper medical treatment."[50]

Waiting for Stevens in Vladivostok was a cable from Daniel Willard informing him of the progress being made of raising and supplying American railroadmen to go to Siberia. They would be known as the Russian Railway Service Corps. In addition, one hundred mechanics were going to be sent to supervise the construction of the engine shops at Vladivostok. Both groups were to be funded by the Russian government through credit from the United

[50]Estep, "John F. Stevens and the Far Eastern Railways," 17.

States. Before they could leave, Washington had to receive a definite Russian agreement on the funding. Stevens was to finalize the arrangements. Also awaiting him were orders from Ambassador Francis to return to Petrograd. England and France were complaining again about the deterioration of the Russian railways and were pressuring the Russian officials to request the United States to send them the best railroad man they had to oversee the railway improvements. British discouragement with Stevens had finally rubbed off on Francis, who had seemed to admire Stevens's leadership but did not appreciate Stevens's assuming authority that belonged to him as ambassador. Francis evidently thought that that British plan for them to settle the Russian railroad problems was the right one. In late September, after more British urging for him to ask Washington to send a more qualified railroad expert to Russia, he sent a cable to Lansing. In it he recommended that the State Department send the "biggest railroad man available, even Willard himself."

While still in Vladivostok, Stevens met with Ustrugov and reaffirmed that the Russians would provide accommodations for American personnel of the Russian Railway Service Corps. He assigned Miller to remain there and assist with their deployment when they arrived. After resting a few days and partially recovering from his illness, Stevens set out once more on the 5,000-mile journey across Siberia back to Petrograd. Darling accompanied him as far as Harbin, where he departed to go home to America by the way of China.[51] Stevens' inspection of the rail lines on the way showed visible evidence that his commission's recom-

---

[51]Catton, "League of Honor," 168, citing Francis to Lansing, Sept. 28, 1917, 861.77/183, RG59, NA.

mendations were beginning to make definite improvements in operation.

His return to Petrograd had been ordered by Washington. When Willard and Lansing had received Francis' request to send "the biggest railroad man available, even Willard himself," they never considered sending a new railroad man to Russia.[52] Stevens had been there three months and had acquired a knowledge of the Russian railways, labor problems, and the personalities in the provisional government. In answer to Francis' request, Willard wired back that Stevens was the best man for the job.

On August 7 President Wilson called a cabinet meeting to discuss the Russian railway situation. He had doubts about sending all of the equipment the Railroad Commission requested. After five weeks of discussions with the Russian Ministry of Transport, Stevens had failed to win approval of his commission's recommendations. If the Russians were unhappy with the Stevens' ministry, they were also unable to provide recommendations of their own. Secretary of the Treasury McAdoo had asked the Russian ambassador in Washington for detailed estimates of their most pressing needs. The ambassador requested a large sum of money but could not provide any exigence for its use.

Wilson's opinion was unsettled. How could the United States send locomotives and more railroad men to Russia when their government refused to send for them and evaded the recommendations of the Stevens mission? He was also disturbed by the attitude of the British, whose ambassador in Petrograd, Sir George Buchanan, had sent a report through their embassy in Washington to the U.S. State Department. He had written that the reason for the

---

[52]Estep, "John F. Stevens and the Far Eastern Railways," 17.

disastrous situation in Russia was the deplorable state into which the equipment of transportation had fallen. Tracks, locomotives, and rolling stock were all worn out. The railway material being considered for shipment from the United States would not be enough for what was needed.

Wilson finally arrived at a decision on the Russian policy. He did not favor the British suggestion of Allied intervention, so now the Railroad Commission, originally intended to explore means of getting American assistance to the eastern front, would assume the function of forestalling Allied interference on the Russian railways. He discussed this with Lansing, and they decided that Stevens should assume an advisory post in Russia's Ministry of Transport, but they didn't want to reveal this to the British officials, who had their own candidate, General A. DeCandoble, in mind to be an advisor. In a vaguely-worded cablegram to Francis so as not to reveal their scheme to the British, Lansing stated that "Stevens would act as director general to the Ministry of Transportation."[53] Then he ordered Francis to have Stevens return from Vladivostok. Francis immediately grasped Lansing's intent that the State Department was trying diplomatically to steer an independent course from Britain and forestall Allied intervention. General DeCandoble didn't know about the plan and assumed Stevens' mission was still limited to its concerns for the Trans-Siberian Railway.

Stevens arrived in Petrograd on October 14, and Ambassador Francis informed him that the State Department had appointed him to what the Russians called "director general" of all the railways but, knowing the Russians, Stevens interpreted it as simply meaning an "advisor." Francis told

---

[53]Catton, "League of Honor," 170–71, citing Lansing to Francis, Oct. 15, 1917, 861.77/197, RG59, NA.

him to keep the appointment a secret. Soon after accept-
ing his new position, he was told that he, Minister of Foreign Affairs Michael I. Tereshchenko, Minister of Ways
and Communications Liverovskii, and Francis were to make
a trip to Mogilev to confer with the army staff in charge
of military railways. After waiting for two weeks for the
Russian officials to begin the trip, Stevens and Francis met
with Tereshchenko and Liverovskii to cancel it and to
request another assignment for Stevens. During the two
weeks of Stevens' inactivity, General Poole of the British
military equipment section in Petrograd goaded Stevens
with a letter questioning Stevens' assessments of the Murman Railway. Poole was the one who had earlier urged the
commission to inspect that line. Poole insinuated that
Stevens had badly underestimated the shortage of locomotives and implied his lower estimates were faulty and
debilitating to the program of American railroad assistance.

Stevens' curt reply was an explanation of his estimates and
reminded Poole that the Railway Commission was only advisory. The U.S. government was not responsible for the
Russian railways, and rumors that the commission planned
to take over operations of the railroads were counter-pro-
ductive and unfounded. The British continued their demands
on Liverovskii and presented to him a comprehensive plan
for Russian railways. It called for the creation of a "Department Relationship" with control under one person instead
of a commission. The department would include a subordinate person for arbitrating labor disputes and who would
oversee problems such as coal supply, reorganization of
engine repair shops, and the improvement of transportation
on waterways. The Russians refused to act on the plan.[54]

---

[54]Catton, "League of Honor," 177.

On October 25, Stevens and Francis met with Tereshchenko and Liverovskii to request again an assignment for Stevens. The engineer suggested that he travel south and implement his commission's recommendations on the railways in the Ukraine. Tereshchenko was reluctant and agitated. Finally he shouted out that his government did not require any assistance and that he was being assailed with "altogether too much advice."[55]

He soon changed his mind, and two days later requested Stevens to inspect the rail line from Moscow to Omsk to see if it were possible to increase food supplies from western Siberia—winter was approaching and a large number of people would be facing starvation. Stevens departed to carry out the assignment. When he arrived in Moscow he found a traffic jam created by 8,000 to 12,000 freight cars loaded with war supplies destined for the front lines. He tried to untangle the obstruction but had little success. He continued eastward to Chelyabinsk, where he arranged for a shipment of some six million bushels of wheat to Moscow and other cities in European Russia.

Before Stevens could finish his trip and complete the task Tereshchenko had given him, a drastic political change took place in Russia. On November 7, 1917, the Bolshevik Revolution occurred in Petrograd. The revolutionaries seized the telegraph office and Winter Palace, then took over the entire city. The next day they organized a new government, which they called the Council of the People's Commissars. Vladimir Lenin was selected chairman, Leon Trotsky as commissar for foreign affairs, and Joseph Stalin as commissar of nationalities. The new government arrested all of the ministers of the provisional government except the

---

[55]Ibid.

Russian Bolshivek, Stevens'
guide before 1917 Revolution.

leader, Alexander Kerensky, who fled the city in disguise
in a car commandeered from the American embassy.[56]

Stevens was returning to Petrograd when he heard about
the Bolshevik coup. His train was halted at Moscow because
rail traffic had been stopped along the tracks from there to
the capital. The Bolsheviks in Moscow were fighting for control of the city, which they eventually achieved. Battles raged
among city streets, and Stevens stayed in his private railway car at the South Station. After several attempts he man-

---

[56]Ibid., p. 176; Arno J. Moyer, *Political Origins of the New Diplomacy, 1917–1918* (New
    Haven: Yale University Press, 1959), 262.

aged to telephone Ambassador Francis, who told Stevens to return to Petrograd. The engineer told him it was impossible, so the ambassador advised him to wait in Moscow until the railroad was restored, and then return.

Stevens had doubts about going back to Petrograd. The Bolsheviks seemed to be gaining the fight in Moscow, and it looked like a civil war. As soon as the railroads were in the hands of the revolutionaries, Stevens implored the commissar of railways to find him a train east. On November 12, his car was attached to a northern train going to Vologda. He waited there two days to see if he could get a train to Petrograd, wanting to cooperate with the ambassador, but there was no train available. Finally, he wired Francis that he was going east and would wait in Harbin for the arrival of the Russian Railway Service Corps from the United States. He was able to get the Bolsheviks to attach his car to the Trans-Siberian Express, the last run of the Express eastward to Vladivostok for more than four years.

His trip lasted nearly two weeks, and he arrived in Harbin on November 24. There he cabled the Washington State Department: "Any further attempts toward helping railroad absolutely useless. . . . I cannot stand Russian winter. Will shortly leave for the United States."[57]

* * *

In Washington, D.C., on July 30, 1917, as civil and military unrest was heating up in Russia, Daniel Willard received Stevens' cablegram requesting military units of railroad men to educate Russians in American operation. He had approved the request and referred it to Samuel Felton, director general of military railways. Felton decided to recruit

---

[57]Estep, "John F. Stevens and the Far Eastern Railways," 17.

the units from northwestern states because, being accustomed to cold winters, they would be better able to withstand the rigors of the Russian climate. Since Felton was also organizing trained railroad men for service in France as part of the U.S. military, Willard assumed that those going to Russia would have the same military status as those going to France and being organized by the same official. He had conferred with the Russian ambassador in the United States, who verified the corps would be paid by the Russian government.[58]

In September Felton met with a group of Russian officials who permitted him to set the salaries of the men in the military railway units, and allowed them to have corresponding military ranks with the officers of the Russian railroads in order to avoid the questions of authorities. Since Russia was paying their salaries, it was necessary to classify the railroad corps as a special unit because all members of the United States military had to be paid by the U.S. government.[59]

A few days later General W. M. Black, chief of engineers of the United States Army and Felton's immediate superior, sent communication to Secretary of War Newton Baker in which he stated:

> I have been informed by higher authority that arrangements have been made by the Russian ambassador for the United States government to send to Russia about 240 railway operating and shop officials for use on the railways in Russia as instructors, and further, that it is considered very desirable that these men have a military rank and grade corresponding to the grades held by similar officials now in Russia employ. Further, that the Russian government has agreed to pay these men such compensation as may

---

[58]Joe M. Feist, "Theirs Not the Reason Why: The Case of the Russian Railway Corps," *Military Affairs* (February 1978): 2.   [59]Ibid.

John Stevens (on train steps) and Col. George Emerson, head of the RRSC.

be deemed fair and equal. . . . It would appear, then, that it will be necessary to appoint these men as officials in a special Corps in a manner similar to that recently adopted for the American Red Cross.[60]

Secretary Baker approved Black's memorandum, and the official formation of the Russian Railway Service Corps was authorized. Using the Red Cross as an example to be followed in forming the RRSC showed that the war department never intended to make the corps a part of the regular army. But the men who volunteered for service in the RRSC never doubted their membership in the military. They made application to the chief of engineers, United States Army, using a form which read: "I have the honor to apply for examination for a commission as (rank) of Engineers in the

[60]Ibid., 2–3.

Officers Reserve Corps, organized under the authority of Congress."[61] Later, the War Department said that because the corps was formed in such a hurry, those forms were unintentionally distributed. Commissions were issued by the War Department and signed by the adjutant general. They read, "You are hereby informed that the President of the United States has appointed you (rank) in the Russian Service Corps."[62] But here, too, there were irregularities. Some of the men questioned the commissions because they did not definitely state that the RRSC was a military organization. They were assured that regular forms would be sent as soon as the crowded conditions of the War Department could be corrected. The doubters were also led to believe that they would be officially sworn in when the proper commissions arrived. Neither action ever happened.

The corps went to Russia and worked on the Siberian railways. They were in that country for nearly a year and a half before they discovered the government's true conception of their status. Before they left the U.S., Felton had informed their leader, George H. Emerson, that the corps was not eligible for the war risk insurance, but assured Emerson that the ruling was probably a technicality that could be corrected. In Siberia during January 1919 Emerson appealed to Stevens because his men were not covered by insurance, they were not receiving pay raises as other U.S. soldiers were, and they were being treated like civilian employees by the armed forces. Stevens cabled the State Department requesting that those problems be remedied. In April the State Department replied that the "RRSC, while organized by the Secretary of War, at the request of the Provisional Government of Russia is, nevertheless, not a part of the United States Army."[63]

---

[61]Ibid., 3.   [62]Ibid.   [63]Ibid., 5.

The corps remained in Siberia until 1920. When the men returned home, they were determined to rectify the situation by gaining full military status for their corps. Between 1920 and 1961, twenty-four bills were introduced in Congress to grant that status. Each time Congress rejected them. Their reasons were that salary payments made to workers in the RRSC came originally from funds of the Russian Embassy, which were made up of loans granted by the U.S. government, and later from funds of the Inter-Allied Technical board, instead of U.S. government funds. Also, Congress was unhappy with the apparently enormous salaries that were paid. They viewed the RRSC as a group of opportunists who were trying to obtain something to which they were not entitled and found their continuous agitation mildly distasteful.[64]

In December 1967, Harry L. Hoskin filed suit in the district court for the District of Columbia against the secretary of the army on behalf of all survivors of the RRSC. The men sought a final judgment declaring that they were entitled to an honorable discharge from the United States Army for service rendered in Siberia. Government lawyers again stressed the fact that the men were paid by the Russian government, and for that reason could not have been in the U.S. military. Presiding Judge Oliver Gasch responded that the provisional government had fallen before the corps arrived, and their salary payments could not have come from "a nonexistent, bankrupt government." The judge declared that if the government sends its funds on a circuitous route, the funds still remain American. His verdict was:

> There is no question here but that these plaintiffs performed the
> military duties to which they were assigned. . . . They take the

---

[64]The opinion of author Joe M. Feist.

position that having been appointed at the direction of the President of the United States by action of the Adjutant General of the Army they are entitled to an Honorable Discharge as a necessary and proper incident to the military service in the absence of any reason for withholding such an Honorable Discharge. The Court agrees.[65]

In 1973 the government appealed the decision, but the United States Court of Appeals upheld judge Gasch's opinion. After fifty-five years of rejection, thirty-three survivors of the RRSC found satisfaction in their victory and received honorable military discharges for their service.

On November 11, 1917, members of the Russian Railway Service Corps, led by Colonel George H. Emerson, departed by train from St. Paul, Minnesota, to begin their service. After a week in San Francisco becoming indoctrinated, receiving medical inoculations, and validating their passports, they boarded the ship *Thomas* on November 19, sailing to Vladivostok.

Five days after the corps' departure from San Francisco, John Stevens was in Harbin. The Bolshevik Revolution had begun in Petrograd and was spreading across Russia. He had decided that he wanted to return home and cabled that information to Willard. The news shocked Washington officials. They had just learned that Stevens was the only one of his original five-man advisory commission who remained in Asia. Miller had been the last to depart, having left Vladivostok early in November.

Willard anxiously cabled Stevens: "Please do not leave Vladivostok until Emerson arrives and until you have heard further from me. . . . Splendid work you have done is greatly appreciated, but it is not finished." Stevens thought about Willard's request and decided he had "no

---

[65]Feist, "Theirs Not the Reason Why," 6.

ambitions other than to accomplish some good," so he wired
Willard that he would wait for the RRSC's arrival and he
would remain in Russia if there were any possibility of
achieving results.

He took a train to Vladivostok, arriving the first of
December, dispirited and failing physically from a return
of blood poisoning that he had had in June. Problems faced
him there. The Soviet revolutionists had gained control over
the city and port facilities, and rumors were circulating that
with Stevens' arrival and the RRSC on its way, the Americans planned to occupy the port and seize the railroad in
order to ensure repayment of Russia's loans. Another rumor
reported that Japan was going to send troops into the city
to oppose the Bolsheviks. In addition, the impending cold
weather might freeze the harbor and prevent the RRSC's ship
from landing.

On December 14, the *Thomas* arrived in the harbor.
Stevens immediately told Colonel Emerson to delay plans
for disembarking because the Bolshevik press was stirring
up the people against the RRSC, and proclaiming they
would not provide accommodations for the 350 men. He
sent a cablegram to Willard stating, "Conditions make it
absolutely imperative to delay decision as to landing for
some time. It may be necessary to sail quick. Cable instantly
placing ship *Thomas* under my orders for any port in Japan
we may select. Lose no time."[66] The situation worsened the
next day when it was learned that a number of the interpreters sent from the U.S. aboard the *Thomas* were radical
Russian exiles. Emerson wouldn't allow the agitators to dis-
embark, but the Bolsheviks demanded their release. Stevens
dispatched another cable to Willard asking for permission

[66]Catton, 198.

to sail to Japan until the situation cleared. Washington must not have understood that the situation was desperate—they sent Colonel Emerson instructions to wait in port until further notice. By then, ice began forming in the harbor. The ice breakers were in the hands o the Bolsheviks, and they couldn't be relied on to keep the harbor open. Without ice breakers, the *Thomas* might become stuck in the harbor.

Stevens discussed the situation with American Consul John K. Caldwell, and both agreed that the *Thomas* must leave right away, despite the orders from Washington. That night Stevens boarded the ship, and the *Thomas* steamed out of the harbor at midnight headed for Japan.[67]

The work of the Advisory Railroad Commission appointed by President Wilson had come to an end. He had sent them to Russia to advise and assist the Russians on their transportation situation and carry out their agenda on whatever railroad matters they might suggest. The commission was to be neither political nor diplomatic, but was to be Wilson's attempt to show support for Russia's new provisional government. Now with the overthrow of that government by the Soviet Council of People's Commissars, which the U.S. would not recognize, Stevens' mission was finished. His commission had made good progress on the Trans-Siberian Railway and played a valuable role in staving off Allied intervention in operating the Russian railroads, which President Wilson had opposed.

---

[67]Ibid., 199.

Stevens solving railroad problems in Siberia
with Japanese official, Nagao.

7

SERVICE WITH THE
INTER-ALLIED TECHNICAL BOARD

Stevens and the Russian Railway Service Corps aboard the *Thomas* arrived in Nagasaki, Japan, on December 19, 1917. They had to hold the ship in the harbor as a floating barracks for three weeks until they could secure sufficient U.S. funds to cover the expenses of the RRSC personnel in hotels ashore.

Stevens continued on to Yokohama, where he became ill once again. Since he was a civilian, the U.S. Naval Hospital there could not admit him as a regular patient. Instead, the commander hosted him in his own quarters until Stevens regained his health. In late January 1918, John Caldwell, American consul at Vladivostok, informed Stevens and Emerson that the RRSC could safely return to Russia. More government authority had been established over mutinous laborers, and Bolshevik soldiers no longer dominated railway workers. The two men went to Vladivostok to survey the situation and then proceeded to Harbin, Manchuria. Stevens began negotiations there with Dmitri Horvath, head of the Chinese Eastern Railway, and L. A. Ustrugov, now retained as a railway official by the new Bolshevik government, to attempt to work out a plan for placing a unit of the RRSC on the Chinese Eastern Railway. The group of Americans would be an entering wedge for the RRSC operations in Siberia.[1] Talks began February 1, and by March 10 Stevens, thinking that he had reached an understanding with the two Russian railroad executives, had one hundred members of the RRSC sent to Harbin. However, five weeks passed before they were able to start working, because Ustrugov kept objecting to when they should begin. Stevens finally warned him that unless the RRSC members were allowed to start work immediately, he would arrange to send the entire contingent back to the U.S. and report to his government that the Russian railway authorities didn't desire assistance from the Americans.[2] The threat worked, and on April 10, 1918, the railroad engineers began performing the work for which they had been recruited. Yet it took four more months before Stevens could bring the remainder of the RRSC members from Japan to Vladivostok to begin work in Russian territory. The long delay was due to the Bolsheviks in Vladivostok who had kept them away, thinking America had sent the railway corps to collect money on a loan which the U.S. had made to Russia. In late June Czechoslovak prisoners of war, who had deserted from the Austrian Army and had arrived in the Far East en route to join French forces on the western front in Europe, overthrew the Bolsheviks in Vladivostok and captured the city.

Corps members were divided into units of fourteen men and located at important division points. They served as advisors to Russian railway officers and directed them in the operation of the various lines. Although some of the Russians were slow and reluctant to adopt new methods, they accomplished many improvements. The establishment of an American system of train-dispatching speeded train

---

[1] Estep, "John F. Stevens and the Far Eastern Railways," 18.   [2] Ibid.

movements and avoided delays. Changes made in repair shops resulted in increased output of locomotives and cars. Introducing sheets of schedules kept operation officers informed about what their trains were doing. Freight tonnage increased and in some cases nearly doubled.[3]

For a long time the Japanese government had wanted to get control of the Chinese Eastern Railroad, an important link in the Siberian system. In January 1918, they had requested permission from Britain and the U.S. to occupy Vladivostok and take control of the Chinese Eastern and Siberian portions of the Trans-Siberian Railways. Their military, the Japanese insisted, would protect those two lines from falling into German possession. Neither the U.S. State Department nor Stevens wanted Japan to be in charge of those railways, because they believed the Russian railroads should not be owned by a foreign nation and recognized that Japan was working toward that end.

Officials in Washington sent Stevens to Tokyo where he, along with American ambassador Roland Morris, began negotiations with Japan's foreign minister, Yasuya Uchida. The result of their discussions was to recommend the establishment of a Japanese-American Commission to supervise operation of all Russian railroads. President Wilson approved the proposal with a suggestion that the commission be international. On October 23, 1918, Japan presented to the Allies another proposition: that all nations with military forces in Siberia create an organization to be known as the Inter-Allied Technical Board, with Stevens as the leader, to supervise operations of the railways in Manchuria and Siberia. Each country would have a single representative on the board.

---

[3]John Frank Stevens Collection, Folder 1, Hoover Institution Archives, Stanford University, Palo Alto, Calif.

Before the Allies could take action on creating a technical board, Germany surrendered on November 11. In a railroad coach in Compiegne Forest at Rethondes, France, both sides signed an armistice that ended World War I. However, in Russia skirmishes and confusion continued on the railway systems. The Allies resumed discussion on Japan's proposal, and in March 1919, representatives from the United States, Japan, Great Britain, China, France, Russia, Italy, and Czechoslovakia voted to organize the Inter-Allied Technical Board with Stevens as president. Its headquarters originally were at Vladivostok, but were later moved to Harbin, where it remained until Stevens disbanded it in 1922.

The board consisted of two committees: military transportation and technical. Ustrugov, minister of ways of communication under Bolshevik authorities at Omsk, was selected to be chairman of the military transportation committee. That committee was responsible for transporting Allied troops still in Siberia. The technical committee was led by Stevens. It was composed of railway experts from Allied countries and had at its disposal the Russian Railway Service Corps of Americans. Actual control of all finances and operation of the Chinese Eastern and Trans-Siberian railroads were its main responsibilities.

Stevens continued his service in Russia for three more years. A lot of his time was spent renovating the Trans-Siberian Railway, which had greatly deteriorated since the Bolsheviks had taken over, and also helping to overcome numerous military and political frictions. On the Baikal section of the Siberian line, Cossack leader Semenoff assumed dictatorship. The rail line's efficiency was reduced due to general unrest among the workers—caused by Semenoff's troops murdering, whipping, and maltreating them—who shortened their labor on the tracks and in repair shops. He traveled in armored trains and appropriated boxcars and coaches for his own use. Japanese troops stationed in that area did not interfere with his activities.[4]

Thousands of Czech troops had been captured during the war and sent to prison camps in Siberia. After the conflict was over, they clamored to be sent home. The U.S. agreed to provide ships at Vladivostok to transport them. Since Stevens' technical committee was responsible for rail movement on those two lines, he had the task of getting 100,000 Czech soldiers out of central Siberia to Vladivostok, over a crippled railway, in the middle of winter. Unfortunately, there were 10,000 Japanese and 8,000 Cossack troops standing between the Czechs and the port. Japan had put soldiers in Siberia to assist the Czechs leaving for home, but they seemed to be doing all they could to prevent them from reaching Vladivostok. Stevens didn't know the reason for Japan's action but surmised they were so afraid of Bolshevism that they wanted to keep the Czechs there for a while as a barrier between themselves and the new Russian government.

Stevens had twenty-five locomotives and five hundred cars moved to Siberia for the evacuation, and Czech soldiers boarded the trains with the Czech leader, General Gaida, in command. However, the trains couldn't move because the Cossack and Japanese troops interfered by seizing locomotives, guarding engine houses, and blowing up a water tank. Stevens sent Colonel B. O. Johnson, who had replaced Colonel Emerson as chairman of the RRSC, to counsel with Cossack Semenoff and the Japanese general. Attempts to

---

[4]Olive Gilbreath, "The Sick Man of Siberia," *Asia* 19, no. 6 (June 19, 1919): 548.

work out a solution failed. Stevens telegraphed Johnson to end the discussions and tell General Gaida that the Inter-Allied Technical Board had done all it could. He would have to fight his way out. Gaida replied that he was reluctant to start a battle with another friendly nation. Stevens informed him, "All right, stay there and starve and freeze."[5]

General Gaida reconsidered his situation and decided to fight. He stationed field artillery batteries, shock troopers, and regular soldiers in front of each train and then announced he was ready to move. Colonel Johnson gave the order to begin. They started and kept on moving. It was a tense situation, as armed Japanese and Cossacks troops had surrounded the depot. If a shot had been fired, the consequences would have been most serious. Nothing happened, and the trains carrying Czech forces arrived safely in Vladivostok.

In 1920, the Russian Railway Corps returned to the U.S., leaving local workers and Allied troops to maintain the railroads. For the next three and a half years (1919–22), Stevens spent his time in active charge of the board's affairs, often having to be a railroad diplomat. Japan kept pushing to gain control of the Chinese Eastern line, owned by Russia and operated by China. Stevens assigned troops of the different nationals to guard various railways. Japan tried to have their troops guard the main line of the Chinese Eastern, but Stevens did not want that to happen, as China would cause trouble. Instead, Stevens gave them the Amur railway, which Japan owned, and the branch line of the Chinese Eastern extending from Harbin to Chauchun where it connected with the South Manchuria. Japan was disgruntled but accepted the situation. He gave Chinese

[5]John Frank Stevens Collection, Stanford University, Stanford, Calif.

troops the task of guarding Chinese Eastern's main line. By diplomatically matching wits, he prevented the Japanese from taking the Chinese Eastern Railroad.

At times Stevens had to deal with problems caused by Bolshevik troops as they advanced across Russia. He forestalled three serious railroad strikes, which would have paralyzed the entire system. He was personally responsible for the use of large funds contributed by the Allies to buy railroad supplies and pay salaries of employees.[6]

When the Inter-Allied Technical Board was organized, the final clause of agreement stipulated that the board would be liquidated when the last foreign soldier left Siberia. The last Japanese soldier sailed from Vladivostok on October 25, 1922. On November 1, Stevens held the last board meeting in Harbin. After attending to important business matters pertaining to their work, the board voted to close. Leaving the secretary, his son, Eugene, to attend to the details, Stevens left Harbin on November 23 to return home.

He had been requested by the Chinese government to be their guest while in China. Chang-tso-lin, Manchurian governor and warlord, also invited him to be his guest as he passed through Mukden on his way south. Stevens spent a day and a half in Manchuria, where he enjoyed a formal dinner in the governor's palace and received valuable gifts of Chinese art. Leaving Mukden on a special train with four young Chinese escorts provided by China, he arrived in Shanghai. There he received the best of accommodations and more gifts and enjoyed visits with American officials. His Chinese hosts took him to their ancient holy city of Hangchow. They carried him seven miles in a chair to show him age-old temples and statues. After two and a half

---

[6]Ibid.

weeks of visiting that country, Stevens boarded the ship *SS President Wilson* to sail to San Francisco. On the way the ship stopped at the ports of Kobe and Yokohama to unload and load freight and passengers. While at Yokohama he went to Tokyo to be the guest of honor at a luncheon given by Foreign Minister Uchida. Stevens' ship continued its voyage, and he arrived in San Francisco on December 27, 1922, having celebrated a pleasant Christmas at sea.

During his railway service in Russia, Stevens held three vital and influential positions: chairman of the American Railway Commission, advisory chairman of America's Russian Railway Service Corps, and president of the Inter-Allied Railway Board. He served under two U.S. presidents, Woodrow Wilson and Warren Harding, and worked with three secretaries of state, Robert Lansing, Bainbridge Colby, and Charles Hughes. Neither president ever expressed to Stevens appreciation for his war service. Alluding to that, he once wrote:

> Both the Wilson and the Harding Administrations supported me wholly—that is they left me completely alone—on my own. During the six years I never received a direct instruction from Washington, nor did I make a written report during that time— until a very short one when I closed out my services.[7]

On his way east from San Francisco, Stevens visited his brother in Denver and his son, Donald, in Ohio. He arrived in New York on January 12, 1923, where he made temporary headquarters. He spent time with another son, John, and his wife, who were visiting from Shanghai, and renewed acquaintance with old friends living in the area. He was informed he had been elected an honorary member of the American Society of Civil Engineers and

---

[7]Ibid.

attended the formal and impressive induction service in New York City. It was a high honor, as at that time honorary membership was limited to fifteen engineers out of a total of eleven thousand.

The Inter-Allied Technical Board records arrived and necessary clerical work regarding the accounts was conducted. On March 16, he submitted his report on general matters concerning the board to the secretary of state. On April 6, he closed the Technical Board accounts, submitted them to the State Department, and was discharged from the war service on April 30, 1923, six years after President Wilson had appointed him in 1917. He was seventy-seven years old.

Secretary of State Hughes released to the press a commendation which he wrote about Stevens:

> I wish to take this occasion to assure you of the high regard in which your work as president of the Inter-Allied Technical Board, and as the American representative thereon, is held by the President as well as by myself and the other members of the government. . . . Your own leading part in this work constitutes a public service of the highest order. I feel that you have contributed much to the well-being of the people of Eastern Siberia and Manchuria and to the early recuperation of their economic life, and that you have advanced the prestige and honor of the United States in that part of the world and with all those who have known of your work.[8]

As the years passed while Stevens was in Russia, he often wearied of his frustrating task and more than once asked to be relieved of his burden, only to yield to State Department pressure to continue in the service of the United States. In January 1922 he had expressed his desire to "throw up at once the Chinese Eastern Railway as a hopeless prob-

[8]Estep, "John F. Stevens and the Far Eastern Railways," 23.

lem." On that occasion, after Secretary of State Hughes personally insisted that his continuance as chairman of the Technical Board was "essential to the preservation of American interests," Stevens subordinated his desires and agreed to continue as the U.S. representative until the Technical Board's activities could be brought to an end.[9]

For his long and trying labors as president of the Technical Board, Stevens was honored by his country with the award of the U.S. Distinguished Service Medal, by France with the Legion of Honor, by Czechoslovakia with its Military Cross and highest civilian decoration, by China with the orders of Chia-Ho and Wen-Hu, and by Japan with the Order of the Rising Sun.

---

[9]Ibid., 23.

8

Later Years' Activities

During the months Stevens waited in New York to be released by the State Department from his military assignment in Russia, he enjoyed visiting his family and acquaintances. He also took pleasure in golfing at St. Andrews Golf Club in Westchester County, where he had been a member for years prior to going to Russia. Knowing he was now retired, a number of groups desired to honor him for his accomplishments and to make use of his consultation abilities.

In 1923, the Baltimore & Ohio Railroad appointed him to be one of their directors. The same year Bates College awarded him the degree of Doctor of Law, which had been conferred on him the previous year in his absence. The honor deeply touched him because he was not a college graduate, and the Bates trustees considered him a credit to the state as one of its native sons.[1]

In 1924, he moved to Southern Pines, North Carolina, where he lived until 1929. Then he moved to back New York for a few months before settling in Baltimore to be near his oldest son, Donald, and his family. While in Southern Pines much of his time was spent on projects and traveling in many parts of the country.[2]

---

[1]Stevens Papers, Lauinger Library, Georgetown University, "Autobiography," box 3, folder 27.　　　　　[2]Ibid.

In September 1925, the people of his hometown, West Gardiner, Maine, wanted to give him a "home coming." One of the affair's leaders was Albert M. Spear, justice of Maine's Supreme Court and a schoolmate of Stevens at the academy they attended. Together they drove to the Grange Hall at French's Corner in West Gardiner where the event was to be held. Nearly two hundred people had assembled for a "genuine Yankee" dinner. Stevens later commented, "How good were the dishes which I knew in my boyhood, and such cooking. In all my wanderings, nothing ever tasted better than did that dinner."[3]

Following the meal, the people gathered in a large hall where Maine's governor, Ralph O. Brewster, and Judge Spear gave short speeches and presented Stevens with a "Resolutions Of Respect." It read in part:

> We, the people of your native town, and of neighboring towns, by this public gathering, embrace this opportunity and this method of expressing to you our sincere appreciation of the signal honor which your eminent career has reflected upon your native town, state and nation, and to manifest our pardonable pride in claiming a larger right than any other community to appropriate the honor of your famous career.[4]

In reflecting on that event, Stevens later wrote,

> Never in the past had I visualized in my mind such a welcome back to the dear old town and its warm hearted people. To feel that one is not forgotten by his old associates, no matter what time or distance intervenes, is a great source of happiness.[5]

In July 1925, the Upper Missouri Historical Association, sponsored by the governors of Minnesota, North Dakota, South Dakota, and Montana, planned a dedication ceremony of a large bronze statue of Stevens. The Great

[3]Ibid.   [4]Ibid.   [5]Ibid.

Stevens at his statue dedication on Marias Pass, Montana.

Northern Railway placed it at the summit of Marias Pass, displaying him as he appeared when he discovered it in 1889. Gaetano Cecere was its sculptor. The railroad and historical association made arrangements for Stevens to attend.

A large number of people were there, some from New York and Portland, Oregon. Among them were Steven's son, Eugene, and his wife; two grandsons, John F. III and Thomas B.; and old friends General W. C. Brown and Alfred Jenkins. Addresses were given by Charles A. Carey, historian of Oregon; Robert Ridgway, president of the American Society of Civil Engineers; and Honorable Pierce Butler, associate justice of the U.S. Supreme Court and a long-time friend. The statue was unveiled by his grandson, John F. III. Stevens expressed his deep appreciation for the honor given him. Music was played by the Great Northern Orchestra. Salutes by western cowboys and congratulations from friends made up a full day for the retired engineer. Ralph Budd, president of Great Northern Railway, gave Stevens a bronze statuette, a replica of the large statue, which Stevens presented to the Maine State Library in Augusta.[6]

On November 8, 1926, Stevens received the following telegram from Queen Marie of Romania while she was visiting the U.S.:

> As I stood on the rear end of the observation car with your devoted friends, John H. Carroll and Stanley Washburn, at Summit, Montana, on the Great Northern Railroad that you built, just as the sun rose over the mountains and struck its first rays the heroic statue of a heroic pioneer, I listened to the story of your great achievement. Stop. From my heart I congratulate you as a great engineer also by suffering, fortitude and faith struggled through these mountain paths to create a highway of empire and shortened the distance between the Atlantic and Pacific by approximately two hundred miles. Stop. I understand that such men as you have made the great western country an empire in itself.[7]

Stevens replied by telegram:

> Your majesty does me great honor and I deeply appreciate the beautiful manner of its expression. Stop. May your reign and life be long.[8]

The Great Northern Railway's original tunnel, which Stevens had built in the Cascade Mountains in 1900, had become intolerable from gas fumes created by the old coal-burning locomotives of early days. In 1909, steam-motive power was abolished, and electric transportation was used

---

[6]Ibid.   [7]Ibid.   [8]Ibid.

through the tunnel due to hydro-electric development in Tumwater Canyon. The railroad then considered a long-standing desire of Jim Hill to have a long, straight tunnel in lieu of the first one. Different routes were studied, and a 17-mile-tunnel plan was formulated, but the cost seemed prohibitive. In 1925, they consulted with Stevens regarding a new tunnel. He knew by heart practically every foot of that Cascade region. After an exhaustive new study, he urged the immediate approval of a 7.79-mile tunnel at an estimated cost of $14 million. On Thanksgiving Day, 1925, Great Northern directors approved his plan, and within two weeks construction was begun and completed in four years.[9]

In March 1925, Stevens was awarded the John Fritz medal, the highest honor that can be bestowed upon a professional person by the united engineering bodies. Candidates who are selected to receive the medal are chosen on behalf of the four national engineering societies of America—Civil, Mining and Metallurgical, Mechanical, and Electrical—by a board of sixteen, four delegates from each society. An affirmative vote of three-fourths is required for selecting the candidate. Some of the former recipients included Alexander G. Bell, Thomas A. Edison, and Alfred Noble. The medal award was established by John Fritz on his eightieth birthday in 1902, and was the first to receive it. Fritz was superintendent of the Cambria and Bethlehem Steel companies and started the Bessemer plant at Bethlehem, where he developed the successful process for making good steel railroad rails from American ores. He was a pioneer in the introduction of the open-hearth process in the U.S., in the manufacturing of Armor plate, and in the hydraulic forging of large shafts. For fifty years his wise

[9]Anderson, *Rails Across the Cascades*, 44–45.

counsel aided in the vast expansion of the American iron and steel industry.[10]

John R. Freeman, chairman of the Committee on Arrangements, presented Stevens to receive the medal. In his remarks he stated,

> . . . in recognition of your distinguished services as a civil engineer, particularly in connection with the Panama Canal, the building of railroads and as administrator of the Chinese Eastern and the Trans-Siberian Railways, I have the honor and the great pleasure to deliver to you this gold medal.[11]

Stevens responded by saying:

> I, as a raw lad, without money, influential friends or technical training, cast my hat into the ring and decided upon an engineering career. Just why I did so, I do not know, . . . but I did and have never regretted it, least of all tonight. . . . I very soon realized that I must overcome as far and as speedily as possible the serious handicap of the lack of a technical education, and so for many weary years, my Bible was the works, written on paper and in timber, iron and stone, of those great men who have long since gone to their reward, who laid so well the foundation of modern American engineering practice. . . . [F]or years my work took me away from petty trumpery of civilization into the wide open spaces and among mountains, where one lives close to Nature, who is the great Comforter and Mother of us all and who is ever ready to whisper a word of cheer to the despondent.[12]

In his paper he later wrote some of his feelings about receiving the John Fritz medal:

> I endeavored in a somewhat incoherent way to express my thanks for the distinguished honor which I had unexpectedly received, but for once in my life my feelings so overcame me that my brain and tongue were nearly useless, . . . the acknowledgment of the service

---

[10]Miles P. Du Val Jr. "John Frank Stevens, Civil Engineer, Explorer, Diplomat and Statesman," Hall of Fame Committee, *American Society Of Civil Engineers,* New York, N.Y., 14–15.   [11]Ibid., 44.   [12]Ibid., 45.

that I performed at Panama was especially dear to me, because I had long since given up any expectation that I would ever be given proper credit, in the public eye, at least, for the work I did there.[13]

Some of Stevens' former overseas acquaintances who could not attend the award presentation sent messages of congratulations:

> I was closely associated with Mr. Stevens in the supervision of the Trans-Siberian and Eastern Railways. . . . May I ask you to be good enough to extend to Mr. Stevens my heartfelt congratulations upon the distinction he is about to receive?
>
> T. Matsudaira
> Japanese Embassy, Washington, D.C.[14]

> I am convinced that when Russia is back on her feet and the people of Russia are in a position to judge and to express their feelings, they will cherish your name as one of Russia's best friends, and will associate your activities with one of the most lofty examples of American spirit.
>
> Boris A. Bokhmeteff
> (former Imperial Russian Ambassador
> to the United States)[15]

> I cannot allow this opportunity to pass without voicing my admiration for the genius of Mr. Stevens as an engineer who has left monuments of his achievements in all parts of the world. A more worthy recipient of the John Fritz gold medal can hardly be named.
>
> Sao Ke Afred Sze
> Chinese Legation—Washington, D.C.[16]

From January 1924 to March 1925, Stevens served as a consultant for Bates & Rogers Construction company on the Eldorado & Santa Fe Railway and also as a financial consultant for T.W. Snow Construction.

---

[13]Stevens Papers, Georgetown University Library, box 3, folder 27.

[14]"John Frank Stevens, Civil Engineer, Explorer, Diplomat, Statesman." John Fritz Medal Board, American Society of Civil Engineers, *Minnesota Historical Society Library,* 48, 51, 49.   [15]Ibid.   [16]Ibid.

In May 1925, he was guest of honor in Buffalo, New York, having served as consulting engineer for the design and specifications of the Peace Bridge across the Niagra River between the United States and Canada. Legislation was enacted by both countries and dedicated to years of friendship between them. Construction began in August 1925 and was finished in August 1927.

Two major engagements took most of Stevens' time in 1926. One took him to Boston where he designed new freight terminals for the Boston and Maine Railroad with its president, George Hanover. For many years the line had been operating four separate terminals, due to the fact that its main lines consisted of the original Boston and Maine (which connected Boston with Portland, Maine), the Old Eastern (whose route was two parallel lines that ran along the north shore between Everett and Lynn), the Boston and Lowell (which connected the two towns for which it was named), and the Fitchburg (a line running from Boston to Troy, New York). The situation was causing delays, duplication in movement of cars, needless expense, and, at, times indifferent service. The new terminal was a large yard that combined into one plant three of the antiquated ones. The fourth yard could be adjusted to conform with the new one whenever expenses justified it. The improvement was an economical investment.[17]

Stevens' other engagement was the reorganization of the Chicago, Milwaukee & St. Paul Railroad. His job required an examination of the main lines for the physical condition of tracks, terminals, equipment, shops, and all of its accessories. Then a report was to be made covering those areas, along with a probable amount of new capital required to put the property in first-class condition.

---

[17]Stevens Papers, Georgetown University, box 3, folder 27.

In January 1927, he was elected president of the American Society of Civil Engineers, an honor for which he had hoped but never expected to receive. Tenure for the president's office was fixed for one year. The society had sections in different parts of the country, forty-four in all. With the society secretary, George T. Seabury, Stevens visited as many of the sections as possible, getting acquainted with members, observing their activities and giving encouragement by their presence and advice.

During that year Stevens visited the Canal Zone, sailing from New York to Cristobal. He stayed there nine days and was greatly impressed by the finished project. He sailed home via Los Angeles, San Diego, and San Francisco, and at all three stops he was entertained by members of the local sections of the American Society of Civil Engineers. After a visit with his brother in Denver, he reached New York in March. In December he delivered a lecture to the faculty and graduating students of the Massachusetts Institute of Technology. His subject was "The Future of the Engineer."

For the first three months of 1928, he spent most of his time in Washington as chairman of an American Society of Civil Engineers committee to attend meetings of the House Flood Committee. It was handling the matter of legislation for the Mississippi River, which had caused a disastrous flood the previous year. The purpose of his committee was to secure the appointment of a joint commission composed of representatives of civilian and Army Corps engineers. That commission was to study the problem of control of the Mississippi flood water, collect and summarize the data, and formulate a comprehensive plan to control the river. The Flood Committee gave Stevens' committee assurance that its plan for a mixed commission of engineers would be favorably reported. However, it adopted a plan submitted by the Army Corps and entirely ignored the engineers in civil life.

Stevens was very disgruntled with the House's action and felt that whenever the Army Corps of Engineers approved or disapproved of a project for which an appropriation of public funds was necessary regarding the Mississippi flood control, Congress members would always accept their plans.[18] Always favoring civilian engineers, Stevens later wrote:

> The ranks of engineers in civil life in this country contain some of the ablest hydraulic engineers in the world, and their opinions and judgment should be consulted in grave problems . . . not only on account of their ability and experience, but also because the great mass of our engineers are taxpayers, and vitally interested in knowing that their money is being spent to the best advantage.[19]

At the commencement of North Carolina University in June 1928, he received the honorary degree of Doctor of Engineering. A week later the University of Michigan awarded him the same honorary degree. Stevens' son, Donald, and his wife were present at both occasions, which made the events even more pleasurable. Not being a technically educated person and receiving two honorary degrees in one week, Stevens appreciated them more deeply.

While in Michigan he received a telegram advising him that the secretary of the interior wanted him to take the chairmanship of a Committee of Engineers to be appointed by Congress. The committee's purpose was to investigate and report on the problem of the Boulder Dam (now the Hoover Dam), proposed to be built on the Colorado River in Arizona. The Colorado River had been flooding for several years and needed to be controlled. The flow had cut a deep channel until the entire river was flowing into the

---

[18]Ibid.   [19]Ibid.

Alamo Channel and the Salton Sink, creating the Salton Sea and often flooding the Imperial Valley in southern California.[20] In December 1928, after several failures, both the House and the Senate had approved the bill and sent it to the president for final approval. President Coolidge signed the bill approving the Boulder Canyon Project in December 1928. (The dam was actually built in Black Canyon rather than Boulder Canyon.)[21] Seven states were involved, including Arizona and California.

Stevens declined the government's request for him to be chairman of a committee of engineers to work on Boulder Dam, because he thought the committee would be handicapped from considering any other plan than the one proponents were advocating. Also, the development of the potential electrical power of the Colorado involved consideration of an important economic problem that the government entirely ignored. In addition, compensation for engineers who were called on to advise was extremely low.[22]

Arizona opposed the construction of the dam because Congress wanted it for producing electrical power, therefore greater financial income, while Arizona needed it for flood control. The Arizona State Colorado River Commission hired Stevens as a consulting engineer to make an independent investigation and prepare a report to assist them in their opposition.[23]

In January 1929, Stevens traveled west to Washington to participate in the ceremony of opening the tunnel through the Cascades that he had designed four years earlier. The

---

[20]Ibid.

[21]Wm. Joe Simonds, *The Boulder Canyon Project—Hoover Dam*, <http://www.usbr.gov/history/hoover.htm>, accessed on July 14, 2001.   [22]Stevens Papers, box 3, folder 27.

[23]John F. Stevens, *The Matter of the Colorado River*, a brief written for the Arizona State Colorado River Commission, Oct. 31, 1928, pp. 3–30, Stanford University Library, Stanford, Calif.

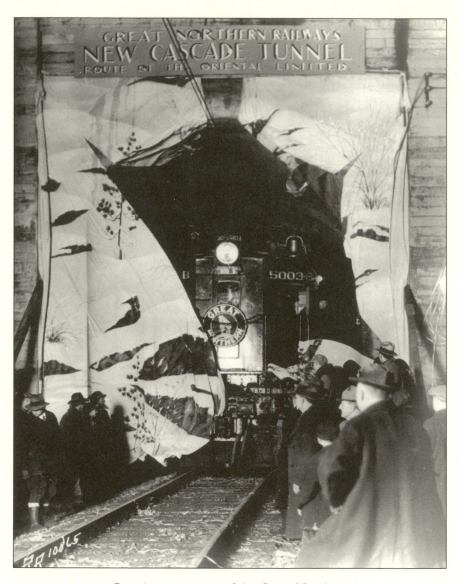

Opening ceremony of the Great Northern
Cascade Tunnel near Leavenworth, Washington.
The Spokesman Review *Newspaper, Spokane, Washington.*

new passageway replaced the first one he had built in 1900. This one had taken three years to build. It was nearly 8 miles long (7.79 miles), the longest in North America and fifth longest in the world. Engineers had so accurately calculated that when the ends met, 3,000 feet underground and four miles from the west portal, they were only seven inches apart, and nine inches different in elevation.[24]

Attending the ceremony along with Stevens were Washington's governor, Roland H. Hartley; Ralph Budd, president of the Great Northern Railway; and other dignitaries. Part of the tunnel-opening program came from the Cascade Mountains, and other portions from New York, Washington, Philadelphia and San Francisco. The entire five-point radio broadcast of the ceremony was synchronized with the running schedule of Great Northern's top train, The Oriental Limited, on its maiden trip through the 8-mile shaft.[25]

Graham McNamee, NBC's noted announcer, was in charge of the activities. Just before the train started through the entrance, president Ralph Budd dedicated the tunnel to the G.N.R.R.'s illustrious founder, James J. Hill. Madame Schumann-Heink, a famous contralto, sang dedicatory songs from San Francisco and popular music was played from New York. The crowd banqueted in the dining hall of the scenic construction camp. In Stevens' honor a huge cake was presented with a replica of a wooded, mountainous summit, complete with trains emerging from the "hole" through the Cascades. In his talk, Stevens described his early explorations in the region that honors him as "Stevens Pass." He spoke of the old tunnel being abandoned and never used again by saying: "Once we hailed that giant-bore as a step in progress. Now I never expect to see that old tunnel again."[26]

---

[24]Anderson, *Rails Across the Cascades*, 45–47.   [25]Ibid.   [26]Ibid.

*   *   *

The annual "Medal Day" meeting of the Franklin Institute was held in the hall of the institute in Philadelphia on May 21, 1930. Their awards program was begun in 1824, when the institute was founded by a group of Philadelphians to train artisans and mechanics in the fundamentals of science. Their awards were at first certificates and later endowed medals. Some of the laureates include Pierre and Marie Curie, Rudolf Diesel, and Albert Einstein. Receiving the medal in 1930 were John F. Stevens and Sir William Henry Bragg, director of the Royal Institution of London.

Stevens' sponsor, Dr. James Barnes, stated in his introduction: "The Panama Canal is without a doubt one of the greatest wonders of the modern world, truly a living memorial to the engineering genius of Dr. Stevens."[27] The president, Nathan Howard, presented the medal and certificate of honorary membership, saying:

> Dr. Stevens, in recognition of the far reaching works in engineering which have just been so well described, the Franklin Institute awards you the Franklin Medal, which is the highest award in its gift. . . . I present to you this medal, and the certificate of Honorary Membership.[28]

Stevens thanked the institute for the honor, and later gave an address on "A Momentous Hour At Panama."

In 1935, Stevens received another great honor by being included in a famous mural painting by William Andrew MacKay in the Roosevelt Memorial Hall, American Museum of Natural History, New York. It shows Chief Engineer Stevens presenting his plan for constructing the Panama Canal to President Theodore Roosevelt, along with

---

[27]Ibid.   [28]Ibid.

Cake served at opening ceremony of tunnel.
The *Spokesman-Review, Spokane, Washington.*

Stevens in retirement.

the engineer's successor, George W. Goethals, and the sanitarian, William C. Gorgas, standing nearby.

Conrad N. Laurer established a trust fund in 1929 to create the Hoover Medal. The award is given to recognize great, unselfish, non-technical services by engineers to humanity. It is held by the American Society of Mechanical Engineers and administered by the Board of Award, which consists of representatives of the American Society of Civil Engineers, American Institute of Mining, Metallurgical and Petroleum

Engineers, American Society of Mechanical Engineers, American Institute of Chemical Engineers and the Institute of Electrical and Electronics Engineers. Herbert Hoover received the first award.[29] In 1938, the medal was presented to Stevens, in his eighty-fifth year.

In 1936, Stevens purchased a home in Southern Pines, where he lived until his death. His closing years brought him painful arthritis, which he patiently endured. He was ninety years old when he died on June 2, 1943, leaving three sons: Eugene Chapin, a commissioner of Southern Pines, Donald French of Baltimore, and John Frank Jr. of Boston. Stevens' family included his wife, Harriet, and five children: son Frank and daughter Abby (both who died in infancy) and the three sons who survived him at his death. Stevens was buried in Boston near his wife, who died in 1917.

John Frank Stevens was a self-made civil engineer whose life was spent in the accomplishment of great engineering feats that surmounted dangerous and difficult obstacles. His career encompassed a large part of the world where he demonstrated his capacity for humanitarian public service beyond his calling. He had been awarded the American Distinguished Service medal, the officer's badge of the Legion of Honor of France, the Military Cross of Czechoslovakia, the order of Chia Ho and Wen Hu of China and the insignia, second class, of Japan's Order of the Rising Sun.[30]

His popularity and fame did not prevent him from practicing his idealistic beliefs. In some of his reminiscings, he wrote:

---

[29]"The Hoover Medal: Recognizing the Best in Engineering," *Newsletter of the Hoover Library Association*, Fall/Winter 2000–2001.

[30]*The Pilot Newspaper of Southern Pines* [North Carolina], vol. 23, no. 26, pp. 1, 8.

. . . an engineer is one, who by faithfully conserving the resources of, and intelligently guiding the forces of nature, makes possible the maximum comfort of the human race. To be such an engineer is to be no mere follower along old and worn out paths, but a dynamic, compelling, economic force that makes people think, and the lack of real thought is the greatest among the several causes that keeps our civilization in its present crude, incomplete and unbalanced social condition.

. . . to be a real engineer he must flavor his methods of work by that great gift to humanity, a controlled, but vivid imagination. He cannot be a mere materialist.[31]

The engineering genius of John Frank Stevens should remain an historical influence, nationally and internationally, in the years to come.

---

[31]Stevens Papers, Georgetown University, box 3, folder 27.

* * *

Bibliography

&

Index

*

BIBLIOGRAPHY

BOOKS

Anderson, Eva. *Across the Cascades*. Wenatchee, Wash.: World Publishing Company, 1952 and 1989.

Bennett, Ira E. *History of the Panama Canal* (Builders Edition). Washington, D.C.: Historical Publishing Company, 1915.

Catton, Theodore. *League of Honor. Woodrow Wilson. and the Stevens Mission To Russia*. Master's Thesis, University of Montana.

Congressional Record, August 1, 1892; March 2, 1905; May 24, 1906; April 25, 1956. Washington, D.C.

Culp, Edwin D. *Stations West*. Caldwell, Idaho: Caxton Printers Ltd., 1972.

Drury, George H. *Historical Guide To North American Railroads*. Bob Hayden, editor. Waukesha, Wisc.: Kalmback Books Publishing Co., 1988.

DuVal, Miles P. Jr. *And The Mountains Will Move*. Palo Alto: Stanford University Press, 1947.

Fahey, John. *Inland Empire—D. C. Corbin and Spokane*. Seattle: University of Washington Press, 1965.

Flandau, Grace, *The Story of Marias Pass*. St. Paul, Minn.: Great Northern Railway Co., 1925.

Gaertner, John T. *North Bank Road. The S.P. & S. Railway*. Pullman: Washington State University Press, 1990.

Hallbrook, Stewart H. *Story of American Railroads*. New York: Crown Publishers, 1947.

Hamilton, James McClellan. *History of Montana*. Portland: Binfords & Mort, 1970.

Martin, Albro. *James J. Hill and the Opening of the Northwest.* New York: Oxford University Press, 1976.

Mayer, Amo J. *Political Origins of the New Diplomacy.* New Haven, Conn.: Yale University Press, 1959.

McCullough, David. *The Path Between The Seas.* New York: Simon and Schuster, 1977.

Morrison, Elting E., ed. *The Letters of Theodore Roosevelt*, vols. 5 and 6. Cambridge, Mass.: Harvard University Press, 1952.

Schapiro, J. Salwyn. *Modem and Contemporary European History.* Cambridge, Mass.: The Riverside Press, 1934.

Schwantes, Carlos A. *Railroad Signatures Across the Northwest.* Seattle: University Of Washington Press, 1993.

Sibert, William L., and John F. Stevens. *The Construction of the Panama Canal.* New York: D. Appleton and Company, 1915.

St. John, Jacqueline D. *John F. Stevens: American Assistance to Russian and Siberian Railroads. 1917–1922.* Master's Thesis, University of Oklahoma, 1984. University Microfilms International, Ann Arbor, Michigan.

Stevens, Hazard. *The Life of Isaac Ingalls Stevens.* Vols. 1 and 2. Boston and New York: Houghton, Mifflin and Company; Cambridge: The Riverside Press, 1901.

Stevens, John F. *An Engineer's Recollections.* New York: McGraw-Hill, Inc., 1935.

Twohy, John Roger. *Ten Spikes To The Rail.* Jenner, Calif.: Goat Rock Publications, 1983.

Ulam, Adam B. *The Bolsheviks.* New York: The McMillan Co., 1965.

Wade, Rex A. *The Russian Search For Peace.* Palo Alto: Stanford University Press, 1969.

Statutes at Large of The United States. Volumes 27, 33 34, Government Printing Office, 1893, 1905 1906.

Josephy, Alvin M. Jr., ed. *American Heritage.* Park Dutchess, NY: American Heritage Publishing Co., Inc., 1964.

PERIODICALS

Brimlow, George F. "John F. Stevens." *Montana Magazine* 3 (Summer 1953): 39–44.

Budd, Ralph. "Northern Rail Lines Across The Divide." *Civil Engineering* (April 1940).

Budd, Ralph. "Address at John Fritz Medal Presentation to John F. Stevens," *Civil Engineering* (March 23, 1935): 18–30.

Chartran, Henry R. "The Call of Siberia to Railway Builders." *Transpacific II* (February 1920).

Colley, Marion T. "Stevens Has Blasted and Bridged His Way Across America." *The American Magazine.*

Carr, John Foster. "The Panama Canal." *The Outlook.* June 2, 1906.

Donovan, Frank Jr. "Canyon War." *Railroad Magazine* 67 (October 1956): 24.

Du Val, Miles P. Jr. "John Frank Stevens, Civil Engineer, Explorer, Diplomat and Statesman." Hall of Fame Committee, *American Society Of Civil Engineers,* New York, N.Y.

Estep, Raymond. "John F. Stevens and the Far Eastern Railways 1917–1923." *Explorers Journal* (March 1970).

Feist, Joe Michael, ed. "Railways and Politics: The Russian Diary Of George Gibbs, 1917." *Wisconsin Magazine of History* (Spring 1979).

Feist, Joe Michael. "Theirs Not The Reason Why—The Case of the Russian Railway Service Corps." *Military Affairs,* no. 42 (February 1978).

Flood, Hon. Daniel J. *Congressional Record,* House. Tuesday, May 29, 1956.

Foord, John. "Siberia And Its Railway." *Asia* (June 1917).

Galvanti, William H. "Recollections of J.F. Stevens and Senator Mitchell." *Oregon Historical Society* 44 (September 1943).

Gardiner, J. B. W. "The Military Position of Russia." *Asia* (June 1917).

Gleason, George. "How the Yanks Are Speeding Up the Longest Railroad in the World." *Independent* 99 (August 16, 1919).

Gilbreath, Olive. "The Sick Man of Siberia." *Asia* 19, no. 6 (June 1919).

"A Great Maine Engineer." Maine Library Bulletin. 979.719ST471G.

"The Great Northern Railway Tunnel Through the Cascades Mts." *Engineering News* (May 18, 1893).

Grenier, Judson A. "A Minnesota Railroad Man in the Far East." *Minnesota History* (September 1963).

Heffelfinger, C. H. "John F. Stevens—A Record of Achievement." *Washington Historical Society* 26, no. 1 (January 1935).

Hidy, Ralph W., and Muriel E. Hidy. "John Frank Stevens, Great Northern Engineer." *Minnesota History* (Winter 1969).

Inkster, Tom H. "John Frank Stevens, American Engineer." *Pacific Northwest Quarterly* 56, no. 2. (April 1965).

Intlekofer, Charles F. "Railroad Construction In Stevens Pass." *The Confluence*, North Central Washington Museum, 9, no. 2 (Summer 1992).

"John Frank Stevens, Civil Engineer, Explorer, Diplomat, Statesman." John Fritz Medal Board, American Society of Civil Engineers, *Minnesota Historical Society Library*.

John Fritz Board of Award. "Presentation of the John Fritz Gold Medal to John Frank Stevens." *Civil Engineering* (March 23, 1925): 8–45.

Long, Robert Crozier. "The Murman Railway Question." *The Nation* 107, no. 2774 (August 31, 1918).

Maltby, Frank B., "In At The Start At Panama, parts 1–4." *Civil Engineering* 15, nos. 6–9 (June–September 1945).

"Origin of the John Fritz Medal," John Ripley Freeman, chairman. Committee of Arrangements, 1925, American Society of Civil Engineers.

Payne, Hon. Frederick G., *Congressional Record,* House. April 25, 1956.

Railway Age Gazette. November 12, 1909; March [c 10], 1910; March 18, 1910; March 22, 1912.

"American Society of Civil Engineers, Stevens' Address on the Panama Canal, at Denver, Colorado." *Transactions*, paper no. 1650. *Civil Engineering* (July [c 20], 1927): 946–67.

Stevens, John F., "Great Northern Railway," *Washington Historical Society* 20, no. 2 (April 1929).

Stevens, John F., "The Matter of the Colorado River," *Stanford University Library*, Stanford, Calif. no. C 719 as (Sept. 8, 1930): 3–30.

"Stevens. John Frank." *The National Cvclopedia.* Vol 32. New York: n.p., 1945.

"Wanted: 3,000 Locomotives."*Literary Digest* (September 1, 1917).

Canadian Pacific Railway, "The History Of the Railways," a pamphlet for tourists. Publisher and date unknown.

Private Papers Collection

"The John F. Stevens Papers," processed by Michael J. North, 1991. Box 1, Folds 26–47; Box 2, Folds 1–29. Special Collection Division, Georgetown University Library, Washington, D.C.

"Stevens Papers." Box 1, Folders 1 and 2. Hoover Institution Archives, Stanford University, Stanford, California.

"Clinton A. Decker Papers." Seeley G. Mudd Manuscript Library, Princeton Unviersity, Princeton, NJ.

Newspapers

Great Falls [Mont.] *Tribune.* July 11, 1991.

Kalispell Daily Inter Lake. July 11, 1991.

The [Southern Pines, NC] *Pilot.* June 4, 1943.

The New York Times. May 4, 1917, to March 10, 1920 (from the Hoover Institute Archives, Stanford, Calif.).

[Spokane, Wash.] *The Spokesman Review.* December 28, 1896; September 23, 1900; September 27, 1905; February 26, 1908; October 26, 1908; November 3 and 21, 1908; June 17, 1909; November 14, 1916.

The Wenatchee [Wash.] *World.* June 23, 1997.

Internet

The Boulder Dam Project, Hoover Dam, Joe Simonds, accessed July 14, 2001. <http://www.usbr.gov/history/hoover.htm>.

"The Cascade Tunnel," Mike's Railway History, accessed July 16, 1996. <http://mikes.railhistory.railfan.net>.

"Deschutes River," June Williams, accessed on July 15, 2000. <http://www.ohwy.com/or/d/desuri.htm>.

Deschutes National Forest. "Deschutes River," R. A. Jense, accessed February 3, 2000. <http://www.fs.fed.us/rb/deschutes/desnf/resource/physical-water/water/rivers/deschutes.html>.

"Deschutes National Forest," R. A. Jensen, "Little Lava Lake," accessed July 15, 2000. <http://www.fs.fed.us/rb/deschutes/desnf/resource/physical-water/lakes/litlava.html>.

"The History of the Franklin Institute Awards and the Bower Awards," Unisys, accessed on July 11, 2001. <http://www.fi.edu/tfi/exhibits/bower/history1.html>.

"The Hoover Medal," Herbert Hoover Presidential Library Association, newsletter, accessed July 11, 2001. <http://www.hooverasso.org/amhoovermedal.htm>.

"Mountain Pass Marias," Bruce Kelly, accessed July 16, 1998. <http://www.railfan.com/a121988.html>.

Peace Bridge Authority—History. "Crossing Paths—Building Futures—The Peace Bridge," accessed July 11, 2001. <http://www.peacebridge.com/history.html>.

INDEX

American Federation of Labor: 86

American Railroad Commission (in Russia): 166, 171, 172, 174, 175, 176, 180, 185, 188, 195, 205; frustrated in efforts, 178; work suspended in revolution, 183

American Society of Civil Engineers: 136, 214, 225

Angus, Richard B: 21

Arango, Ricardo, 110

Arizona State Colorado River Commission: 227

Army Corps of Engineers: 104, 127, 130

Bailey Gatzert (steamboat) : 147

Baker, Newton D: 160, 161, 187, 199, 200

Balboa, Vasco Nunez de: 77

Baldwin Locomotive: 169

Baltimore & Ohio Railroad, 160, 217

Baltimore: 217

Bangs, Anson M: 122-123

Bastidas, Rodrigo de: 76-77

Bates & Rogers Construction: 223

Beckler, Elbridge: 46, 47, 48, 51, 55, 62–63, 64

Belding, W. M: 103

Bellingham, Wash: 58, 59

Bend, Oreg: 143, 145, 152, 153, 155

Bierd, W. G: 132, 133, 141

Big Hill: 33

Bitterroot River: 50

Black, Lt. Col. Edward R: 111

Black, W. M: 199

Blackfeet Indians: 48, 49, 52

Blue Mountains: 143

Bolich, D. W: 109

Bolsheviks: 181, 182, 196, 197, 198, 204, 205, 207, 208

Boschke, George W: 147, 149

Boston & Maine Railroad: 139, 224

Bow River: 30

Brewster, Ralph O: 218

British Columbia: 22

Brooke, George: 110

Brooke, Lt. Mark: 80, 111

Browning, Mont: 52, 55

Brusilov, Alexis: 180

John Frank Stevens: American Trailblazer
by Odin Baugh
has been produced in an edition of 750 copies.

✳   ✳   ✳

The typeface used is Caslon.
Design by Ariane C. Smith
under the direction of Robert A. Clark.
Printing by Thomson-Shore, Inc.,
of Dexter, Michigan.

✳